THE
GEOGRAPHY
OF THE
CANADIAN
NORTH

THE
GEOGRAPHY
OF THE
CANADIAN
NORTH

Issues and Challenges

Robert M. Bone

TORONTO
OXFORD UNIVERSITY PRESS
1992

Oxford University Press, 70 Wynford Drive, Don Mills, Ontario M3C 1J9

Toronto Oxford New York
Delhi Bombay Calcutta Madras Karachi Petaling Jaya
Singapore Hong Kong Tokyo Nairobi Dar es Salaam Cape Town
Melbourne Aukland

and associated companies in
Berlin Ibadan

Canadian Cataloguing in Publication Data
Bone, Robert M.
The geography of the Canadian north

Includes bibliographical references and index.
ISBN 0-19-540772-5

1. Canada, Northern–Geography. 2. Canada, Northern–Economic conditions.
3. Natural resources–Canada, Northern. 4. Indians of North America–Canada,
Northern. I. Title.

HC117.N67B65 1992   917.19   C91-095109-8

Design by Heather Delfino

Contents

List of Figures

List of Tables

Preface

Over the last fifty years profound changes have been creating a new Canadian North. They include the emergence of the resource industry as the dominant economy, Native self-government and land claims as the main political issue, and industrial pollution as the major environmental problem. The unleashing of such powerful economic, social, and political forces is almost unprecedented in Canada. They have altered the definition of northern development from its almost exclusive association with economic growth to the much broader issues of the so-called hidden — environmental and social — costs of development, and the place of Native people in the northern economy.

'Free trade' has ensured the continuation of the hinterland role for the Canadian North. While the Arctic is just beginning to experience resource development, the Subarctic, particularly the northern areas of provinces, is almost fully integrated into the world economy. Within each northern province, the subarctic zone is treated as a hinterland serving the interests of the province. The Native people and the lands they have used in the past as hunters and trappers have become enmeshed in the resource economy. Native people have not been full partners in the process of development, and have often been marginalized by the weakening of their hunting/trapping economy. Native adults' low rate of participation in the wage economy is a complex subject, worthy of a much fuller discussion than found in this book. One explanation is that the recent shift from a land-based economy to a settlement one has not allowed Native adults sufficient time to obtain the educational and job skills necessary for wage employment; another is that most Natives live in remote communities where there are few employment or business opportunities. There are as well fundamental cultural differences between Natives and non-Natives. Native culture stresses collectivism, while Canadian culture emphasizes individualism — a cultural trait critical for the market economy and a dominant factor in the workplace. This and other cultural differences inhibit some Native adults from seeking and holding jobs in the wage economy. Comprehensive land claims settlements such as the James Bay Agreement have gone some distance to rectifying the imbalance by allowing Native-controlled regional economies, but there are major trade-offs. The Cree residents of northern Quebec strongly oppose the current plans of Quebec Hydro to 'develop' another part of the James Bay river system (the Great Whale Project), because they will lose valuable hunting lands, making their land-based life-style still more difficult to pursue. Not all resource projects have the same level of environmental and social implications as hydroelectric development. The Norman Wells Oil Expansion and Pipeline Project, for instance, had relatively little environmental or social impact.

The reader of this book will note (and I hope agree) that northern development is perceived differently by most Native and non-Native Canadians. The first mentioned see the North as a 'homeland'; others see it as a 'hinterland'. Yet the high natural rate of increase of the Native population, the slow rate of economic growth in Native communities, the increasing use of southern-based workers to fly into remote mines and construction sites, and other issues tend to transcend these two perspectives. The existence of a 'population trap' in many Native communities, and the massive economic leakage of wages to southern Canada, have lasting implications for the North and its regional development.

Teaching courses on the regional geography of the North, and conducting seminars on Northern development revealed a serious gap. While there are many articles or book chapters on parts of polar Canada, a regional textbook dealing with both the territorial and provincial norths did not yet exist. It is precisely in the hope of rectifying this situation that the present book has been conceived and structured. Many have helped in this endeavour. Richard Teleky, formerly of Oxford University Press, identified the lacuna in the geographic literature and encouraged me to write a regional text. Sheila Meldrum of the Department of Indian Affairs and Northern Development read and commented on an early draft, while my colleagues at the University of Saskatchewan helped in many ways, including supplying references, commenting on chapters, and generally being supportive. Professors Barr, Chakravarti, McConnell, Randall, Smith, Tough, Waiser, and Wilson deserve special mention as do Professor Ironside and Dr Weissling from the University of Alberta's Geography Department. Regional geography texts often deal with many themes and the art of balancing these themes is most challenging. This book is no exception. The comments of an anonymous reader of an early draft led to a more extensive discussion of environmental issues. Many of my graduate students, perhaps unknowingly, helped formulate and temper basic concepts in this book as they took part in seminar debates and revisions of their theses. Field work and travel too played a key role, both as a means of keeping in touch with northerners and the realities of the North and as a confirmation of its varied nature and complex character. My days at the now defunct Institute for Northern Studies at the University of Saskatchewan were most rewarding and the opportunity to talk with those from other disciplines and northern places was a delight. The librarians at the Circumpolar Institute Library at the University of Alberta and our own librarians in Government Documents made library research a most pleasant experience. Olive Koyama, editor with Oxford University Press, has made the text much more readable than my copy, and Keith Bigelow's fine maps and graphs have much enhanced the book. The initial draft of the text was prepared in early 1990 when I enjoyed a half Sabbatical Leave from the University of Saskatchewan. During this hectic time and the following months of revision, I have been deeply grateful to my wife Karen for her understanding and her willingness to shoulder so many of my familial responsibilities.

To my family: Karen, Mike, Chris and Signe

I

SETTING THE STAGE

Chapter 1

NORTHERN PERCEPTIONS

Canadians hold a number of visions of the North. Two dominate current thinking: the northern frontier, and the northern homeland. The northern frontier image is of a place where people are pitted against a harsh environment that contains great wealth. This is a popular perception among southern Canadians. The northern frontier image has its roots in the fur trade and its lineage stretches back to the early European explorers who sought to find an ocean route to the riches of the Orient. History gives us exciting tales of early explorers, the adventures of the coureurs de bois, and prospectors' dreams of finding an Eldorado. A more modern version of the image is that of a resource frontier where fabulously rich resources are ripe for development and where large-scale resource projects generate rapid economic growth, thereby solving many of the North's economic problems and adding to Canada's wealth. The notion of a rich resource frontier has deep historic roots, going back to the day of Cartier, Frobisher, and Hudson. The northern frontier, as a source of vast wealth, was given more credence by the Klondike gold strike in the late 19th century. In the current century, the search for northern riches is not confined to valuable minerals like gold and silver but includes base metals such as lead and zinc, timber, and water power. In this sense, the northern frontier image has been modified over time but still involves the notion of the North as a place of yet-to-be-discovered wealth.

Northerners, particularly Native northerners, hold a homeland image of the North. While Canadians think of themselves as a northern people, those living in the North have a special, deeper commitment to that place. Hamelin (1979: 9) described this feeling as 'a trait as deeply anchored as a European's attachment to the site of his hamlet or his valley'. In this single statement, Hamelin has captured the geographer's notion of place and the parallel idea of regional consciousness. Regional consciousness is an appreciation of local natural features, cultural traits, and economic issues. It is the basis of commonality which provides a distinctive regional 'personality' such as is found in the Maritimes and the Prairies. Often, it evokes a commitment to regional interests over national ones. This aspect of regionalism may manifest itself as a political force. It gains strength from real and perceived injustices, many of which appear to stem from actions taken by those living outside of the region, particularly those politicians living in Ottawa. One such northern issue is the desire by the two territorial governments for provincial status and the reluctance of the federal government and provinces to deal with this issue.

For those Canadians born and raised in the North, regional consciousness is based on the concept of 'homeland'. By living, working, and playing in a northern environment, they have developed a deep and lasting attachment to their surroundings. In much of the North, these Canadians are also of aboriginal descent. With their long history in the North, their land-based economy, and aboriginal land claims, there is a decidedly 'Native' theme to the northern consciousness. Through exposure of land claim issues in the media, Canadians have awakened to the realities of a different North and the urgent need to resolve these outstanding claims. This awakening began in 1976 with the extensive television coverage of the Mackenzie Valley Pipeline Inquiry (Berger 1977). This inquiry, conducted by Thomas Berger, considered the social, environmental, and economic impact of a proposed gas pipeline project which would connect Prudhoe Bay gas reserves with markets in the United States. In meetings held in northern communities, environmental and social concerns were given a different perspective–a Native perspective. Canadians were surprised to find that industrial development was so strongly opposed by the Dene and that there were 'unresolved' land claims. The media have continued their northern and Native coverage and so reshaped Canadians' image of the North and of themselves. Unresolved Native land claims, industrial pollution of northern lands and waters, and inequalities between the northern hinterland and the rest of Canada now form part of that image and, in turn, are redefining the Canadian identity (Page 1986: 23; Zaslow 1988: 387).

Defining the North

Geographically, the North is not an easily defined entity. Its southern boundary is marked by the places where the Subarctic gives way to other natural environments such as the grasslands of the Prairies. More precisely, this physical boundary follows the southern extent of the northern boreal forest, with two major caveats. Firstly, the southern half of the mountain forest in British Columbia and Alberta is excluded. Secondly, the black spruce forests of the Gaspé and much of Newfoundland are also excluded.

The actual boundary is really a transition zone. As in all natural regions, a set of common physical characteristics exists. These characteristics are well displayed within the 'core' of the region, but toward its outer limits they tend to blend with the characteristics of neighbouring regions. Such a transition zone is made more complex and its spatial expression less precise by changes in elevation and relief. Higher elevation results in cooler temperatures, and mountain-type relief can result in heavy rainfall on western slopes but a rain-shadow effect on eastern slopes. The resulting changes in vegetation can be significant. As well, sudden changes in relief can alter vegetative patterns, making the precise demarcation of a natural vegetative boundary extremely difficult if not impossible.

Within the North, therefore, there is a wide variety of physical conditions. Yet this complex natural world, when simplified into two major geographic regions called the Arctic and the Subarctic, effectively defines the North. The extent of this region is enormous, covering nearly 80% of the land and water bodies making up Canada.

The Arctic (land, water, and ice) extends over an area of about 3 million square kilometres. Most of this natural region is found in the Northwest Territories, though it also appears in Newfoundland, Quebec, Manitoba, and Yukon. The Arctic has no summer. Its land surface is very young in geological time because much of its land mass emerged from beneath the Laurentide Ice Sheet only a few thousand years ago. Because its soils are frozen for most of the year, they are very immature and can only support quick-maturing and shallow-rooted plants. The resulting tundra vegetation often covers only a small portion of the land, leaving a bare surface exposed.

The Subarctic, stretching in a broad belt across Canada, is by far the largest natural region in Canada, encompassing nearly 4.5 million square kilometres. Unlike the Arctic, this region has a distinct summer period which allows a forest vegetation to flourish. The boreal forest covers much of this region. Coniferous trees, including spruces, pines, and firs, are common. Within this boreal landscape, wetlands are widespread.

The human geography of these two natural regions varies sharply. The Subarctic offers more settlement possibilities and is relatively densely settled with a population (Native and non-Native) of nearly 1.5 million. The Arctic, on the other hand, remains a homeland for Inuit peoples and contains fewer than 30,000 people. Vast areas in the Arctic Archipelago and the Barren Lands are uninhabited.

Defining the North in terms of natural regions has one serious draw-

Figure 1.1
The North

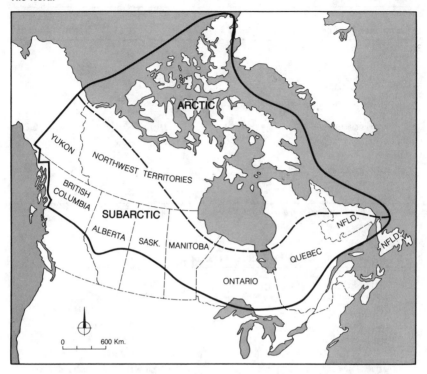

The Canadian North consists of Yukon, the Northwest Territories, and the northern areas of seven provinces. It is also divided into two natural zones, the Arctic and the Subarctic. It is part of the Circumpolar World, and its minority Native population is often perceived as belonging to the Fourth World.

back. Statistical information is prepared for administrative units and is not readily available by natural regions. For this reason, demographic, economic, and social data derived from censuses necessitate a conversion of the North to a set of census divisions that best correspond with the natural North. A discussion of the 29 census divisions selected for this purpose is found in the chapter on population.

Size and Character

The sheer size of the North often surprises Canadians, particularly those who live in southern Canada. Except for Alberta and British Columbia, the northern areas of the remaining seven provinces comprise at least half of their total territory. Of all the provinces, Quebec has the largest 'northern' area, making up 81% of its territory. All of the territories lie in the North. The Northwest Territories and Yukon Territory contain 3.9

Table 1.1
Approximate Land and Freshwater Areas (000 km²)

Province or Territory	Total Area	Northern Area	Percentages North	Canada
Newfoundland	406	300	74	3.0
Yukon	484	484	100	4.9
Manitoba	650	480	74	4.6
Saskatchewan	652	325	50	3.3
Alberta	661	310	47	3.1
British Columbia	948	375	40	3.8
Ontario	1,069	700	65	7.0
Quebec	1,541	1,250	81	12.6
Northwest Territories	3,426	3,426	100	34.4
Canada	9,971	7,650	100	76.7

SOURCE: Adapted from *Canada Year Book 1988*: 1–16; and Hamelin 1988a: 26.

Table 1.2
Common Characteristics of the Political Areas of the North

* remoteness	* sparse population
* cold environment	* urban population
* permafrost	* high proportion of Natives
* resource-based economy	* unsettled land claims
* export of resources	* dependency on government
* import of manufactured goods	* imported foods
* external ownership of companies	* core/periphery structure

million square kilometres of land and freshwater, while the northern provinces have approximately 3.7 million square kilometres. Together, the territories and northern sectors of the provinces occupy just over three-quarters of Canada (Table 1.1).

Within these northern lands, a number of common characteristics underlie the settlement, transportation, and resource development pattern found in each political region. Because of these common features, the issues and challenges facing Canadians living in different parts of the North are similar. These characteristics are listed in Table 1.2.

Cold Environment

Perhaps the most recognizable characteristic is the cold environment. While all Canadians are familiar with winter, those in the North face a truly cold climate. Williams (1986: 6–7) defines this region as having a negative annual heat balance and therefore an area with permanently frozen ground. In some places, permafrost extends to several hundred metres below the surface of the ground. One measure of 'coldness' is

Figure 1.2
The Cold Environment

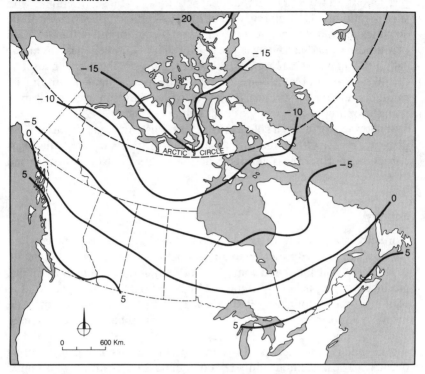

Each isotherm on the map indicates a mean annual air temperature in degrees Celsius. The Canadian North is considered to encompass those areas with a mean annual air temperature below 0°C. Beyond the Arctic Circle, mean annual air temperatures are usually less than −10°C, and permafrost is common. Adapted from Ripley 1987: 155.

offered by the mean annual air temperatures for the North. These temperatures also illustrate the spatial variation of this 'coldness' (Figure 1.2).

Another characteristic of the cold environment is the length of its winter nights. Areas north of the Arctic Circle are in total darkness for part of the winter (see Figure 1.2). This means that residents of high latitude communities are subjected to one or more days of total darkness. Even within the 'zone of darkness', there is considerable variation in the number of days in which the sun does not rise above the horizon. At Alert (82° 29′), there are nearly five months of darkness while at Inuvik (68° 21′), there is just over one month of total darkness. The reverse conditions occur in the summer; Alert has nearly five months of continuous sunlight and Inuvik has just over one month of continuous sunlight–providing there is no cloudy weather.

There are economic, social, and psychological implications of this 'darkness' phenomenon for southern Canadians. For those living in high latitudes, the continuous darkness of the arctic winter imposes severe restrictions on work and recreational activities for those unfamiliar with such conditions. The combination of darkness and a reduction of normal activities may adversely affect the emotional equilibrium and motivation of southern Canadians living in the North. This phenomenon is commonly called 'bushed'. The reappearance of the sun on the horizon usually provokes a favourable response from 'transplanted' Canadians. In Inuvik, residents celebrate the return of the sun with a Sunrise Festival featuring a huge bonfire and a variety of community events!

While the North has a cold environment, this environment varies from place to place. Other features of the North also vary spatially. This concept of northernness and its variation across the North is expressed by the term, nordicity.

Nordicity

Nordicity is a measure of the degree of 'northernness' of a place. This concept was created by a Canadian geographer, Louis-Edmond Hamelin (1972, 1979, and 1988) and combines both physical and human elements. Nordicity accomplishes two things: (1) it provides a quantitative definition of the southern boundary of the North, and (2) it allows a composite measure of northernness for any place. Nordicity is based on ten selected variables which are supposed to represent all facets of the North. The sum of 'polar' values assigned to each variable results in a measure of its nordicity. The maximum number of polar units, 1,000, can only be attained at the North Pole. The 200 polar-unit isoline represents the southern limit of the Canadian North. In 1977, Hamelin calculated that Isachsen had the highest measure of nordicity of any community, with 925 polar units, while Vancouver had only 35. Settlements in the Arctic Archipelago tend to have the highest ratings (Alert, at 878 and Eureka, at 857), though Resolute with its superior transportation facilities and larger population has a lower nordicity rating of 775. The nordicity ratings for other centres are shown in Table 1.3 and the location of these centres can be found on Figure 1.3. The method for calculating polar values is found in Appendix 1 (page 242). In 1975, another version of nordicity was produced (Burns, Richardson, and Hall 1975: 41–3).

Hamelin divided Canada into five zones, two representing southern Canada and three, northern Canada. Southern Canada consists of two zones, Principal Ecumene and Near North (secondary ecumene). These two zones are often referred to as Base Canada. In the North, there are three zones: Middle North, Far North, and Extreme North. The southern edge of the Middle North occurs when the polar units equal 200. This imaginary line lies just north of the agricultural lands of Canada and the transcontinental route of the Canadian National Railway (see Figure 1.3). The northern boundary of the Middle North is defined as 500 polar units

Table 1.3
Nordicity Values for Selected Canadian Centres

Centre	Polar Units	Centre	Polar Units
Halifax	43	Thompson	258
Montreal	45	Fort Nelson	282
Timmins	67	Whitehorse	283
Calgary	94	Schefferville	295
Winnipeg	111	Uranium City	396
St John's	115	Kuujjuarapik	414
Edmonton	125	Aklavik	511
Chibougamau	151	Iqaluit	584
The Pas	185	Old Crow	624
Grande Prairie	198	Sachs Harbour	764

SOURCE: Hamelin 1979: 71–4.

while the Far North spans the area between 500 and 800 polar units. The Extreme North lies beyond 800 polar units. In essence, the Middle North is equivalent to the Subarctic (although the more 'populated' sector of the Subarctic is allocated to southern Canada). The Arctic is divided into the Far North and the Extreme North. The Extreme North includes most of the Arctic Archipelago while the Far North includes Baffin Island, northern Quebec, and the Barren Lands of the Northwest Territories.

Other versions of the boundaries separating these regions have been published by Hamelin. In 1972, a map depicted a simplified southern boundary and a northward shift in the polar boundary of the Middle North along the lower Mackenzie Valley and in the Ungava Peninsula (Hamelin 1972: 34). The most recent version is found in the *Canadian Encyclopedia* (Hamelin 1988a: 1505).

Hamelin is describing the North from a 'southern' perspective which reflects attitudes, beliefs, and values held by people residing in that part of Canada. Their image of the North is quite different from that of those living in the North. For instance, each descriptive label–Middle, Far, and Extreme norths–may seem strange and out of place to a Canadian born north of the 60th Parallel. Such a Canadian might have a different mental map of Canada with the North described as the 'centre' and southern Canada as the 'distant land'. Usher (1987: 527) has criticized the concept because it presents a southern perception of the North.

Yet, for this very reason, Hamelin's nordicity concept has a practical use. Hamelin (1972b) used his nordicity index to develop a zonal system of isolation allowances for federal workers. The vast majority of these employees were orginally from southern Canada. For them, living in a small Native community as an administrator, nurse, or teacher was equivalent to living in the Third World and represented a sacrifice warranting additional pay and staff housing. The allowances were also a

Figure 1.3
Canadian Nordicity, *circa* **1975**

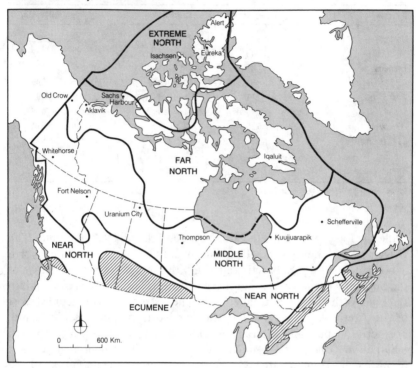

Adapted from Hamelin 1975: 150.

Vignette 1.1   The Dynamic Nature of Nordicity

One interesting feature of nordicity is that the value for a particular place can change over time. Hamelin (1979: 35) states that Saskatoon had a polar value of over 200 in 1881 and, because of changes in its population size and its central place in the Canadian transportation system, it now is considered a 'southern' place with a nordicity rating of 116. Similarly, Hamelin describes the evolution of nordicity at Chibougamau as around 400 in the 1880s while, in 1979, he gave it a value of 151. At that time, he speculated that, given the role Chibougamau is expected to play in the James Bay Project, its nordicity could drop below 100.

---

means of luring professional people to work in isolated northern communities.

The two territorial governments employ a similar system to establish 'living allowances' which make up the higher cost of living in isolated places, and the federal government allows northern residents an income tax reduction (the Northern Residents Deduction). As with most spatial systems, there is a problem with 'boundaries'. In 1989, a federal task force recommended that the boundary line for the federal government's northern and isolated tax benefit program be set at the 57th Parallel in British Columbia. The effect of this move would remove many northern British Columbia taxpayers from continuing to receive these tax benefits. In 1988, an average wage earner in Mackenzie, B.C. would pay about $5000 less federal income tax because of the northern tax deductions (*Globe and Mail* 1989). In 1990 the federal government revised this program by introducing two zones: the northern, and the intermediate. Eligible tax payers in the northern zone qualified for full 1990 tax benefits, while those in the intermediate zone received 50% of these benefits. Residents of Mackenzie, in the northern zone, continue to be eligible for all northern tax deductions.

The northern and isolated tax benefit program reflects economic differences between northern and southern Canada. These differences reflect the high cost of living in the North and the desire for amenities found in large, southern cities.

Economic Variation in the North

The economic landscape of northern Canada follows a north-south orientation, reflected in the Canadian transportation network linking parts of the North with adjacent areas of southern Canada. In this sense, the transportation network identifies a number of economic 'Norths' whereby large, southern cities have extended their trading areas into the North. The dominance of metropolitan centres is greatest in the territorial North and in those provincial Norths that do not have a large regional centre. The extent of these trading areas varies. Thunder Bay, for instance, services smaller communities in northwestern Ontario while Winnipeg's trading influence stretches beyond the 60th Parallel into the arctic settlements situated along the western shoreline of Hudson Bay. In remote areas, such as northern Saskatchewan, local companies and residents purchase many of their goods and services in the south because they are not available in the small northern stores. The economic consequences of this trading relationship are twofold: the south benefits from northern spending patterns, and these spending patterns slow the development of a business sector and more employment opportunities in the North.

Dual Economies

Each of northern Canada's two peoples, Native and non-Native, have a distinct economy system deeply rooted in their cultural aspirations, goals, and values. These two economies may be described as the resource economy, and the Native economy. The resource economy is the dominant commercial force in the North. It produces primary products for export to world markets. Multinational companies and Crown corporations play a leading role in this resource-based economy because vast sums of capital are required. The seemingly endless demand for more and more raw materials and energy ties the northern resource economy more and more closely with the world economy. The North's main export market is the United States. The 1989 Free Trade Agreement with the United States is expected to increase the flow of raw materials and energy from the North to our southern neighbours.

The Native economy has evolved from a subsistence hunting economy to one in which wage employment, transfer payments, and trapping provide cash income while hunting provides country food. While this economy is unable to match the high levels of per capita output of the resource economy, it does encompass Native traditional activities and values. Since contact with Europeans, there has been a link between the Native economy and the external economies, first the French economy, then the British, and now the Canadian economy. Originally, this link was forged by the fur trade in which Natives bartered furs and wild food for manufactured goods and food stuffs. Over time, the needs and wants of Native Canadians have grown and, as with other Canadians, wage employment offers the main avenue to the necessary income to obtain the desired goods and services, whether it be a snowmobile or a case of beer.

Core / Periphery Model

Northern development takes place within the framework of the world economy. The core/periphery model (sometimes referred to as the heartland/hinterland theory) describes the economic relationship between the resource economy of the North and the world market economy. It reinforces the northern frontier image: a frontier hinterland functioning in a larger economic system. The industrialized core dominates events and controls the pace of resource development in the resource hinterland. This asymmetrical relationship is manifested through four key factors: decision-making, investment, technical innovations, and migration. The first three factors are governed by profit-making criteria and controlled by institutions in the core. The last factor, migration, occurs in response to job opportunities. In this model, core regions require mostly raw materials and energy, and this demand triggers economic development in the peripheral areas.

Homeland or Hinterland

Geographers, including Lotz 1970, Hamelin 1972, Usher 1982 and 1987, Bone and Green 1987, and Ironside 1990, and other scholars (Morton 1972, Cruikshank 1977, Berger 1977, Dacks 1981, Diubaldo 1984, Asch 1984, Gamble 1986, Salisbury 1986, Grant 1988, and Stabler 1989) have

Vignette 1.2 Development or Underdevelopment?

Initially, hinterland development concentrates on resource development, while at a later stage the hinterland's economy may become more diversified and therefore more mature (Friedmann 1972). Some geographers and economists suggest that the developed core receives most benefits from this relationship, leaving the hinterland worse off than before (Myrdal 1957 and de Souza 1986). Some go further, claiming that this relationship is exploiting peripheral areas and eventually leads to a state of underdevelopment of indigenous peoples (Frank 1969, Muller-Wille and Pelto 1979, Dyck 1985, Pretes 1988, and Weissling 1989).

presented a wide variety of mental images of the North. Two perspectives–homeland and hinterland–now dominate Canadian thought and scientific writing. The notion of a conflict between the two images of development was first articulated by Berger (1977) and then expanded upon by Dacks (1981), Usher (1982), and Asch (1984). While there are many shades of meaning attached to each image, interaction between the two models has given a new depth to the image of northern development.

The hinterland model is largely subscribed to by non-northerners who envision the North as a place needing 'development'. Such development often means exploiting the natural resources of the North, and so creating wealth for the developers, their workers, and others providing services and goods to the resource project. Northerners not directly involved would benefit from improved access to goods and services from southern Canada. For example, the construction of a northern highway would reduce the cost of shipping foodstuffs from the south to polar communities.

The hinterland model assumes that the northern economy will be integrated more closely into Canadian economic life and into the world economic system, and that persons are rewarded according to their role in this process. This development process is well under way as the market economy has now spread to all corners of the North, seeking and exploiting mineral wealth, forest products, and energy. A second component of the hinterland model is the export of the products of the resource economy to markets around the world. The liberalization of trade between countries, and particularly the Free Trade Agreement with the United States, is expected to accelerate this trend. A third assumption is that the regional population is homogeneous. The model does not recognize substantial cultural differences and major ethnic socio-economic variations within regional populations, a serious weakness when applied to northern Canada, where Native Canadians form a substantial element of the population.

The hinterland model can be interpreted as the advance of Canadian industrial society into the North, displacing the existing Native economy

(Rowley 1978: 73). Four implications are: (1) a 'dual' economy exists in the North; (2) the role of the resource economy is to supply raw materials and energy to the world economy; (3) external decision-making controls much of its economic destiny; and (4) the North is integrated into the global economy. For example, a decline in oil prices in the late 1980s caused oil companies to delay their plans to exploit the Beaufort Sea oil and gas deposits. This link between world prices for resources and the rate of resource development in the North is a fundamental factor in hinterland development.

Canadians living in the North but particularly those of Native ancestry consider the North a homeland. Each is concerned about common economic and political issues but from a different perspective. Non-Native northerners, secure in their place in Canadian society, wish to see the North receive more economic benefits from resource development and more political control over their lives. Yukoners have long argued for provincial status as a means of gaining more control over their affairs. On the other hand, Native northerners are concerned with other issues, especially the impact of resource development on hunting and trapping, land claims, and Native self-government. Resource development does present economic attractions to Native people in the form of jobs and business opportunities, but there is also an uneasy feeling that this type of development leads only to short-term construction jobs and, with a failure to generate permanent jobs, a greater dependency on welfare. Resource development, by adversely affecting wildlife habitat and by drawing people away from their traditional activities, may weaken the cohesiveness of Native society.

Political Variations Within the North

The Canadian North's two territories and northern areas of seven provinces have created nine 'political' Norths. Each province and territory has considerable control over economic events taking place within its boundaries, and therefore these boundaries have real meaning for development. Corporate decisions, based on economic considerations, are influenced by public subsidies and tax concessions. Provinces have more political powers to shape northern development because they have responsibility for natural resources, whereas in the territories the federal government still holds this power. The Northern Accord (1988) represents a first stop in transferring power to the territories. It promises to give the territorial governments control over oil and gas resource development.

At another level, a shift of power from the federal government to Native people is also affecting northern development. As Native people move toward self-government, their role in the process of development is expected to increase. In the Northwest Territories, Native leaders have proposed that province-like native governments be created as part of their land claims settlement. Existing Native governments tend to function only at the band or Métis council level and have little or no control over resource development beyond their settlements. The major exceptions are the result of two major land claims agreements (James Bay and Northern Quebec 1975, and Inuvialuit, 1984). Both resulted in regional Native governments. These Native governments have the organizational structure, capital resources, and political mandate to engage in economic, political, and social development. Beyond this level of regional government, the next step is the creation of a 'province-like' government. The land claim of Inuit Canadians living in the arctic lands of the Northwest Territories might lead to the formation of such a new political state within Canada where an Inuit-controlled territorial government could be formed.

Issues and Challenges

The North is in a state of transition. Many issues and challenges need to be resolved, but there are three in particular. These are:

(1) the effect of the resource economy on the northern economy and its people, (2) the impact of this economy on the environment, and (3) the place of Native people in the resource economy.

Since they are interrelated, each affects the other. This linkage forms the crux of this book. Given the nature of the North, these issues may be resolved differently in the Subarctic than in the Arctic. The next two chapters provide brief sketches of the physical geography and the history of the North. This background information will equip the reader to better understand the nature and process of northern development. The main thrust of the text is the matter of northern development and its impact on the environment and Native people, although the reader is forewarned that all too often the environment has been ravaged and Native people placed in a state of under-development.

The framework for this book is anchored in four ideas. The first is that the human geography found in the North can best be understood within the context of two regions–the Arctic and Subarctic. The second is that the nature and direction of northern geography is best understood by the process of development. In the early history of the North, development was associated with the fur trade; now it is dominated by the resource industry, supported, in turn, by all levels of government. Next, economic development is controlled by the relationship between the world economy and the northern one, best expressed by the core/periphery model. In this model, events are initiated and controlled by the core. Lastly, the existence of two peoples in the North warrants special attention. Each society is engaged in 'their' economy, though Native people often seek wage employment to supplement their subsistence production. This dualistic

North can best be described as 'frontier dualism'. Frontier dualism occurs where two cultures share the same territory: one culture is engaged in an export-oriented commercial economy, while the other one is more concerned with subsistence and land-based activities. Both are locked into the global economy and are evolving in response to global economic forces and the impact of land-claims settlements.

Unlike cultural dualism, however, frontier dualism does not predict a particular outcome, that is, the disappearance of the Native economy. Frontier dualism assumes that peaceful and fair solutions will continue to result from land-claims negotiating. Such resolutions will accelerate the changing nature and goals of northern development, balancing the aspirations of both peoples. Given the limited renewable resource base in the North, its small local market, and the long distances to major suppliers and world markets, this process will not be an easy one. These economic weaknesses underlie many of the issues and challenges found in the North and discussed in the following chapters.

Selected Readings

Berger, Thomas R., 1977. *Northern Frontier, Northern Homeland: The Report of the Mackenzie Valley Pipeline Inquiry*. Ottawa: Minister of Supply and Services.

Bone, Robert M. and Milford Green, 1987. 'Frontier Dualism in Northern Saskatchewan', *The Operational Geographer* 12: 21–4.

Hamelin, Louis-Edmond, 1979. *Canadian Nordicity: It's Your North Too*. Translated by William Barr. Montreal: Harvest House Ltd.

Page, Robert, 1986. *Northern Development: The Canadian Dilemma*. Toronto: McClelland and Stewart Ltd.

Pretes, Michael, 1987. 'Underdevelopment in Two Norths: The Brazilian Amazon and the Canadian Arctic', *Arctic* 41 (2): 109–16.

Stabler, Jack C., 1989. 'Dualism and Development in the Northwest Territories', *Economic Development and Cultural Change* 37 (4): 805–40.

Usher, Peter J, 1987. 'The North: One Land, Two Ways of Life', in McCann, L.D., *Heartland and Hinterland: A Geography of Canada*, 2nd ed. Scarborough: Prentice-Hall.

Weissling, Lee E., 1989. 'Arctic Canada and Zambia: A Comparison of Development Processes in the Fourth and Third Worlds', *Arctic* 42 (3): 208–16.

Chapter 2

THE PHYSICAL BASE

Arctic and Subarctic Environments

Across the northern half of North America, there are two natural regions–
the Arctic and the Subarctic. Climate has played a key role in forming
these two biomes. The arctic climate, for instance, is associated with a
particular type of natural vegetation, soils, and biological life. All these
natural features are interrelated. Their natural vegetation provides the
most visible spatial pattern (Figure 2.1). These same natural regions form
the cultural homelands of the Inuit and Indian peoples.

   The common denominator found in both regions, but particularly in
the Arctic, is the presence of a cold environment, which results in slow
growth of biological life, natural vegetation, and soil formation. Because
of the polar climate, the presence of permafrost, cold ocean bodies, and
the chilling effect of .he Labrador current, only the continental areas of
the Subarctic experience warm weather for any length of time. Besides
being a cold environment, most of the North receives so little precipitation
that it is termed a cold desert. These two factors, low temperatures and
low precipitation, make the North among the least productive biological
areas of the world (Figure 2.2). At the same time, such a slow rate of
plant and biological life means that the natural environment is slow to
recover from natural or anthropogenic damage. The fragile nature of the

Figure 2.1
Arctic and Subarctic Natural Vegetation Zones

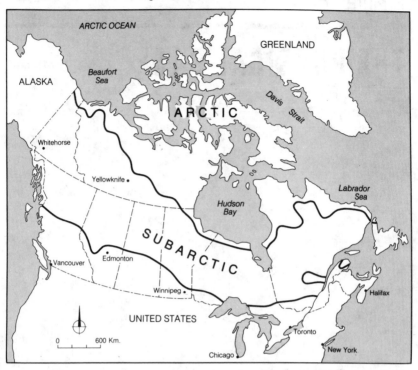

The natural divisions of the Canadian North have both economic and cultural implications. The great boreal forest of the Subarctic supports many of the region's economic activities. In the Arctic, the ancestors of the Inuit developed a technology and social system necessary for a successful hunting economy. Adapted from Trenhaile 1990: 22–3.

Figure 2.2
The Productivity of Natural Vegetation

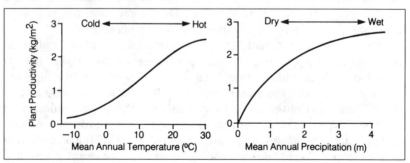

Plant growth varies with mean annual temperatures and precipitation. A combination of low temperatures and meagre precipitation results in the very low levels of plant growth, and hence limited soil development and biological life, typical of the Arctic. Adapted from Haggett 1983: 72.

polar environments is indicated by the relationship of temperature and precipitation to plant growth (Figure 2.2).

The Arctic has the coldest environment in Canada. It lies north of the treeline, found primarily in the Northwest Territories but also in Quebec, Newfoundland, Ontario, Yukon, and Manitoba. It is characterized by a very cold climate where the warmest month has a mean monthly temperature of less than 10°C. Under such climatic conditions, normal tree growth is not possible and a tundra vegetation is formed. True zonal soils are not found in the Arctic. In many places, rock and unconsolidated gravels and sands form the ground surface. Where thin soils have formed in the continuous permafrost zone, they are often classified as cryosols (soils which have an active or thawed layer of less than one metre in depth). Under these conditions, soil-forming processes work extremely slowly because soil temperatures are often just above the freezing point. Even when summer air temperatures thaw the top few centimetres of the ground, the presence of permafrost beneath this thin active layer acts not only as a cooling agent but also as a barrier to water.

Arctic vegetation is often divided into two zones–Low Arctic and High Arctic. The Low Arctic occupies the mainland while the High Arctic is found in parts of Keewatin and the Arctic Archipelago. The Low Arctic zone is characterized by nearly complete plant cover, including many shrubs, sedges, and scrub trees such as birch and willow. Tussock sedge and low tundra shrubs are the summer grazing of barren ground caribou, which are still a major source of food for Native people. Heath, herbs, and lichen are the typical plants. While trees such as willows do grow, they reach a height of only a few centimetres.

In the High Arctic zone, freezing temperatures occur almost daily and permafrost is present near the surface throughout the year so that even tundra vegetation growth is inhibited. Periglacial features–bare rock and a variety of unconsolidated materials–are commonly found. Such barren arctic lowlands are called polar deserts (Vignette 2.1).

The treeline often serves as the boundary between the Arctic and the Subarctic. Ecologically, the treeline represents a major 'break' between the two regions, though in fact the 'break' is a gradual one consisting of a wooded tundra (Vignette 2.2). The proportion of tundra to forest varies but, toward the Arctic zone, only a few stunted trees are found in the more protected areas of river valleys. Spruce and larch are the predominant species found in the wooded tundra. This transition zone also represents a biological and cultural boundary. Many subarctic animals, including the beaver and moose, are not found in the Arctic and of the aboriginal peoples only the Inuit developed the technology permitting year-round life in the Arctic.

The Subarctic, the largest natural region in North America, is covered by the boreal forest in a continuous belt from Newfoundland to the Rocky Mountains. Its summers are short but warm, allowing for a much richer

Vignette 2.1   Polar Desert

The term 'polar desert' usually describes barren areas of bare rock, shattered bedrock, and sterile gravel. These conditions are essentially an extreme form of fell field, with vegetation cover, at least of higher plants, reduced to near zero. Actually, this initial impression of barrenness is usually inaccurate. It is hard to find a square metre of polar desert that has no form of vegetation growing on it. By far the most important group of plants here are the lichens.

Lichens (pronounced *like'-unz*) are the foot soldiers or unsung heroes of the tundra environment. Lichens are in a sense a group of primitive plants, although it is more correct to think of them as a colony or association of two kinds of plants living together. The main body of a lichen is called a thallus; it consists of a dense, tangled mass of threadlike fungal material. This fungal matrix varies greatly in shape from one kind of lichen to another, but each kind has a characteristic form by which it can usually be identified. Encapsulated with the fungal strands are quantities of single-celled algae. These tiny green plants supply the colony with nutrients.

SOURCE: Young 1989: 198.

Vignette 2.2   The Treeline

The treeline represents a dividing zone between the Arctic and Subarctic regions. It closely corresponds to the 10°C monthly mean temperature for July. Other natural factors, such as the depth of the active layer of permafrost, protection from wind, south slope radiation, and well drained land, may result in patches of trees growing north of this isotherm, or conversely, tundra occurring south of it. Then too, weather varies from year to year, causing the annual position of the isotherm to fluctuate somewhat. During the relatively warm decade of the 1980s, for instance, summer temperatures were much higher in high latitudes than in most previous decades. All of these factors have produced a transition zone of wooded tundra between the Arctic and Subarctic regions.

Three centres close to this imaginary line are Aklavik (68°12′N), Churchill (58° 48′N), and Cartwright (53°36′N). Aklavik, located on a deltaic island at the mouth of the Mackenzie River, has a warmest-month average mean temperature of 14°C; Churchill, situated near the shore of Hudson Bay, has an average July temperature of 11.8°C; and Cartwright, lying along the Labrador Coast, has a mean July figure of 12°. About 250 km inland from the coast, temperatures are much higher and, at Happy Valley-Goose Bay, the mean July temperature is 15.8°. The continental effect of this large land mass coupled with the summer cooling caused by the ocean waters of Hudson Bay have resulted in a north-west to south-east trend to the 'treeline'.

vegetation cover than that found in the Arctic. Coniferous trees predominate. Variation in this forest allows for the identification of four subzones: wooded tundra, lichen woodland, closed boreal forest, and forest parkland. The wooded tundra forms the transition zone beyond which lies the tundra. Patches of spruce and larch are found in sheltered, low-lying areas while high, more exposed lands are treeless. The lichen woodland is a savannah-like area consisting of a thick groundcover of lichens interspersed with stands of spruce and pine. The closed boreal forest, a dense forest of fir, spruce, and pine, is found in southern Yukon, a small part of the Northwest Territories (principally the upper Mackenzie Valley), and the seven provinces. Within this zone, the forest cover, is broken by a variety of wetlands including muskeg and peat bogs. In this wet environment, black spruce and larch are the most common species. On the better drained land, species of spruce, fir, pine, and larch are common along with stands of poplar and birch. Towards its southern limits, broadleaf trees, particularly aspen and birch, are found. The forest parkland, a narrow transition area adjacent to the Canadian Prairies, is a combination of forest and mid-latitude grasslands. Small bushes, including blueberries, and grasses form the floor cover.

Infertile podzolic soils are common in the Subarctic. These thin, acidic soils are best formed under cool, wet growing conditions where the principal vegetative litter is derived from a coniferous forest. A low evaporation rate, immature drainage, and permafrost ensure an excess of ground moisture, resulting in severely leached soils and the widespread occurrence of marsh and bogs. In spite of these environmental handicaps, soil formation is much more active in the Subarctic than it is in the Arctic. One reason is that the longer summer temperatures allow the ground to thaw to a depth of several metres, promoting biological activity and chemical action. Another is that the vegetation cover is much denser in the Subarctic than in the Arctic, thus providing much more plant litter for soil formation.

Subarctic lakes and forests provide an attractive environment for many animals and wildfowl. Unlike animals in the Arctic, wildlife in the Subarctic often hibernates during the winter months. Within this forested biome, a wide variety of animals make their home, including the beaver, moose, and grouse. The abundance of lakes, marsh, and muskeg provides a favourable habitat for many waterfowl and big game animals such as moose which graze on shoreline vegetation. In the winter, caribou herds migrate from the Arctic into the northern forest.

Polar Climate

The cold nature of the North has shaped the vegetation, soils, and biological life found in Arctic and Subarctic. This cold environment–the polar climate–results primarily from low receipts of solar energy. It is this limited solar energy entering the northern hemisphere which creates the two types of polar climate: the arctic and the subarctic.

Polar climates lie in high latitudes. They are characterized by low levels of solar energy and precipitation. On average, the earth's poles receive 40% less radiation than the equator (Lawford 1988: 144). At the Arctic Circle, solar radiation is reduced to zero for one day and, at higher latitudes, for longer periods of times, reaching six months at the North Pole. Such low levels of radiation result in continuous cooling of the land and the build-up of masses of frigid arctic air which are associated with daily high temperatures of − 40°C or lower. These frigid arctic air masses often surge southwards, causing blizzard conditions in the Canadian Prairies, subzero temperatures in eastern Canada and freezing temperatures in the southern United States. In the spring, solar radiation increases but, because of the reflective snow-cover, much of its effect is lost, with up to 80% of the solar radiation being reflected. Until the snow cover melts, the breaking of winter's grip is slowed. Once the snow is gone, temperatures recover rapidly and the ensuing warm weather quickly melts the ice from lakes, rivers, and the sea coast. By early July, sea ice has disappeared from Hudson Bay and, a few weeks later, from along the edge of the arctic sea coast. Polar Pack Ice, of course, continues to occupy most of the Arctic Ocean. During the short summer, massive amounts of solar radiation reach the northern lands and waters. With long daylight hours, the landscape warms and plants quickly appear and flower. Daily summer temperatures in the Mackenzie Valley and southern Yukon can reach into the low 30°C. By September, however, cooling begins and winter soon regains its grip on the polar lands and waters.

The seasonal pattern of air temperatures is illustrated in Figure 2.3. The winter regime is shown by the mean daily temperature for January, demonstrating its cold continental nature. The coldest January temperatures are found in two places–the north end of Ellesmere and Axel Heiberg islands and just south of Boothia Peninsula. On the other hand, the mean daily temperature for July demonstrates both the warming effect of the northern land mass and the cooling effect of the Arctic Ocean (both open water and permanent ice pack), Hudson Bay, and the Labrador Current. There is a marked northwest to southeast trend in the July isotherms with a warm 'corridor' extending down the Mackenzie Valley.

Precipitation in the North is generally light with the least amount falling in the Arctic (Figure 2.4). The main reasons are the inability of cold air to absorb moisture and the existence of ice on northern lakes, Hudson Bay, and the Arctic Ocean for much of the year. For these reasons, it is not surprising that the lowest annual precipitation occurs in the ice-locked islands situated in the Arctic Archipelago. Victoria Island, for example, has such scant rain and snowfall (often less than 140 mm annually) that it is described as an 'arctic desert'. Resolute Bay, which is located just 16°

Figure 2.3
Mean Daily Temperatures in Degrees Centigrade, January and July

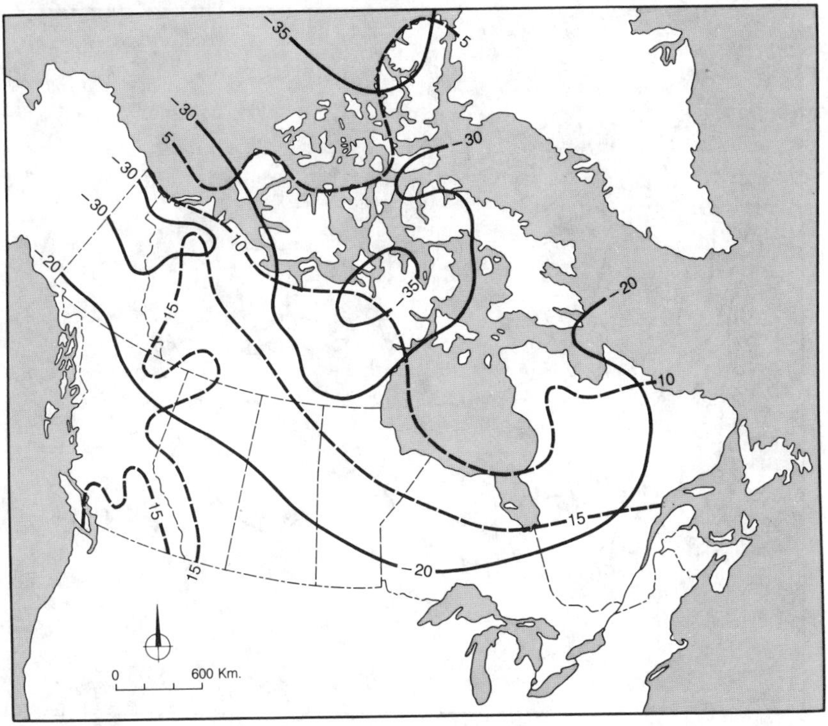

July isotherms indicate a northwest-southeast trend, with a 'warm' corridor extending into the Mackenzie River valley. The minus 20°C January isotherm almost reaches Lake Superior.
SOURCE: Hare and Thomas 1979: 37.

south of the North Pole, receives less than 140 mm per year. Further south, at the northern edge of the Subarctic, precipitation is heavier. At Yellowknife, the annual precipitation is around 250 mm. The Subarctic, on the other hand, generally receives more precipitation, usually over 300 mm annually. Precipitation does vary, however. A few areas do receive much greater precipitation. The greatest amount of precipitation occurs in the Cordillera and along the Atlantic coast. In these two areas, the high terrain results in orographic precipitation (rain or snow caused when warm, moisture-laden air is forced to rise over hills or mountains and is cooled in the process). The south coast of Baffin Island receives around 400 mm annually while the figures for the southern coast of Labrador and the mid-North of Quebec exceed 800 mm. Most precipitation falls as snow. In the spring, rivers and streams are swollen by melting snow and ice. Some communities along the Mackenzie River are subject to spring flooding. Fort Simpson, situated on an island at the confluence of the Liard and Mackenzie rivers, has been inundated a number of times, and

Figure 2.4
Mean Annual Precipitation (in millimetres)

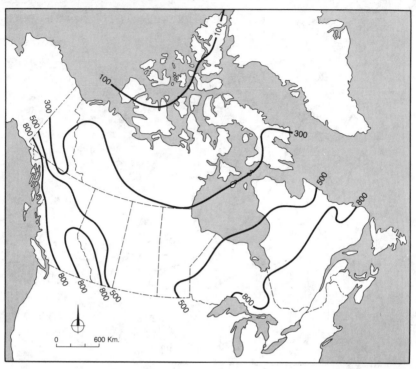

Two cyclonic systems affect Canadian northern precipitation. Eastward-moving storms originating in the Pacific Ocean generate much of the precipitation in the Subarctic. Summer cyclonic disturbances from the Atlantic Ocean and Hudson Bay produce much of the precipitation on northern Quebec, Labrador, and southern Baffin Island. Annual variations in precipitation for this region have affected the output of the James Bay and Churchill Falls hydroelectric projects.
SOURCE: Hare and Thomas 1979: 41.

Aklavik, located in the delta of the Mackenzie River, is threatened by spring flood waters almost every year. The occurrence of spring flooding at Aklavik is so regular that, in the late 1950s, the federal government decided to create the new town of Inuvik rather than expand the community of Aklavik.

In addition to the Arctic and Subarctic climatic types, most climatic maps also depict a Cordillera climate, a composite of several climates. In this text, the mountainous area of northern British Columbia, Yukon, and Northwest Territories is treated as part of the Subarctic. Elevation and topography have important roles in determining the nature of the Cordillera climate. Another climatic type found in the North, the Ice Cap climate, is associated with large glaciers, ice caps, and icefields, and has

Table 2.1
Mean Monthly Temperatures for Chibougamau, Iqaluit, Prince George, and Resolute

Centre	Jan	Feb	Mar	Apr	May	June
Chibougamau	− 18.4	− 10.5	− 1.0	6.4	13.3	15.8
Iqaluit	− 25.6	− 25.9	− 22.7	− 14.3	− 3.2	3.4
Prince George	− 12.1	− 6.1	− 1.8	4.3	9.3	12.9
Resolute	− 32.1	− 33.2	− 31.4	− 23.1	− 10.9	− 0.6

	July	Aug	Sept	Oct	Nov	Dec	Year
Chibougamau	15.8	14.1	9.1	2.6	− 4.7	− 15.9	− 0.8
Iqaluit	7.6	6.9	2.4	− 5.0	− 13.0	− 21.8	− 9.3
Prince George	15.1	14.1	9.7	4.8	− 2.9	− 7.9	3.3
Resolute	4.1	2.4	− 5.1	− 15.1	− 24.5	− 29.3	− 16.6

SOURCE: Environment Canada, 1982. *Canadian Climatic Normals*, Table 3.

a mean temperature below freezing for all months. Figure 2.1 shows the distribution of arctic and subarctic climate types.

The arctic climate is defined as one in which the average mean temperature for the warmest month is less than 10°C. It is distinguished by extremely long winters and a brief, cool summer. It lies north of the treeline and includes all of the Arctic Archipelago, the coastal zone stretching from the Beaufort Sea to the coast of Labrador, and much of the interior of the Northern Territories known as the 'Barren Lands'. While the arctic climate is normally associated with high latitudes, it does reach into the middle latitudes along the Labrador Coast. Here, the chilling effect of the cold Labrador Current keeps summer temperatures along the east coast of Canada low, allowing the arctic climate to extend southwards to the Labrador coast and the northern tip of Newfoundland. Resolute and Iqaliut have arctic climates and their mean monthly temperature regimes are shown in Table 2.1.

Unlike its arctic counterpart, the subarctic climate has a distinct but short, warm summer. Normally, this climate is found in continental locations, and is characterized by a wide range in temperatures, from extremely cold winter temperatures to hot summer temperatures. This continental effect results in record daily cold temperatures being set in Yukon rather than in the Arctic Archipelago. The coldest temperature recorded in Canada (− 62.8°C) was at Snag. In contrast to the cold winter temperatures, a number of 'hot' summer days can occur. In the summer of 1989, for example, during a 'heat wave' in the Mackenzie Valley, Norman Wells recorded an all-time daily high of 35°C.

Winter is the dominant season and, although an occasional summer day may be extremely hot, summers in the Subarctic are short, usually less than three months. Freezing temperatures can occur at any time. By late August, cool fall-like weather is common and there is an impending

Figure 2.5
Climatic Regions of Canada

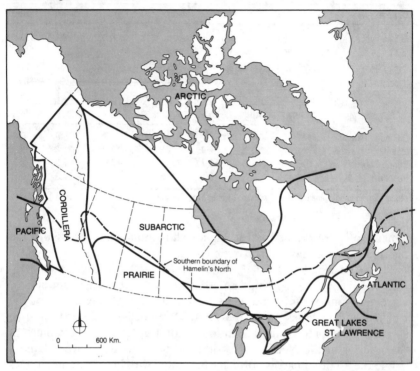

After Hare and Thomas 1979: 17.

sense of winter. By November, the land, lakes, and rivers are frozen, and the dominant season, winter, persists until late April or May. During the winter, the warming influence of the Pacific keeps temperatures of the western Subarctic relatively high while the frozen Hudson Bay reinforces the cold continental effect on the winter temperatures of the eastern Subarctic. The average January temperatures for Prince George and Chibougamau reflect these differing climatic influences (Table 2.1). The reverse situation occurs in the summer when Chibougamau has higher average monthly temperatures than Prince George.

Geomorphic Regions

Across the Canadian North, the surficial geology forms the basis of major geomorphic regions. Such regions must have three major characteristics–they cover a large, contiguous area with similar relief features, their landforms have been shaped, for the most part, by similar geomorphic processes, and each landform has had a similar geological history and

Vignette 2.3 The Subarctic Climate in the Cordillera

There are many factors which control climate, such as latitude, topography, the proximity of bodies of water, and the nature of the underlying surface. In Yukon, due to its mountainous nature, topography becomes very important. The territory benefits from Pacific airflows from the west, while high mountain ranges block Arctic air masses from the north. The mountain ranges also affect atmospheric circulation patterns, the amount, frequency and type of precipitation, winds, atmospheric pressure, and the local radiation regime. Elevation in particular plays a major role in determining temperature. During the winter, a strong surface-based inversion develops in Yukon due to the net negative radiation balance. Thus, temperatures tend to increase with height, especially in the bottom 1500 metres of the atmosphere.

SOURCE: Adapted from Etkin 1989:12

possesses a common geological structure. These macro-regions should be seen easily from a high flying aircraft or from satellite photographs.

Over the past 30 years, physical scientists, including Bostock 1964, Bird 1972, Graf 1987, and Slaymaker 1988, have described a set of geomorphic regions for Canada. The North's five geomorphic regions have been identified: the Canadian Shield, the Interior Plains, the Cordillera, the Hudson Bay Lowland, and the Arctic Lands (Figure 2.6). Geologically speaking, the land surface of these five regions is young, and repeated glaciation and deglaciation have had a profound impact in shaping landforms and determining drainage patterns (Heginbottom 1989: 576).

The Canadian Shield, stretching from Labrador to the Northwest Territories, is the largest geomorphic region in the Canadian North. It is the geological core of the northern landscape. Its Pre-Cambrian rocks are more than 2.5 billion years old and they are found under most of the geologically more recently formed strata, such as the Interior Plains and the Hudson Bay Lowland.

Over most of the northern areas of Manitoba, Ontario, Quebec, Labrador, the northern half of Saskatchewan, the northeast corner of Alberta and about half of the Northwest Territories the Canadian Shield is exposed at the earth's surface. Shaped like a saucer, its highest elevations are found around its outer limits while the central area lies beneath the waters of Hudson Bay. Much of the exposed Canadian Shield consists of a rough, rolling upland. Along the east coast from the Labrador to Baffin Island, the Shield has been strongly uplifted and is deeply riven by fjords, creating spectacular scenery.

The Interior Plains, flat to gently rolling landscape, lie between the Cordillera and the Canadian Shield. Their sedimentary rocks were formed after the end of the Pre-Cambrian era (some 1/2 billion years ago) and include Cretaceous-age rocks formed some 100 million years ago. At the

Figure 2.6
Geomorphic Regions of Canada

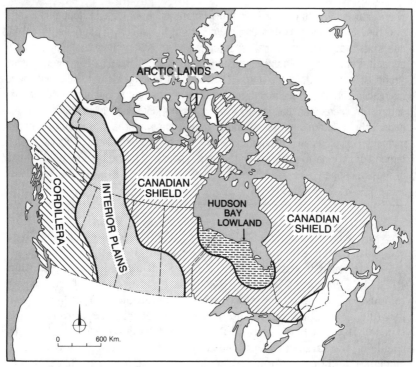

SOURCE: After Bird and Slaymaker 1988.

end of the last glacial advance, these sedimentary strata were covered by a mantle of glacially deposited debris. In places, glacial lakes were formed. Silts and clays were deposited. Later, when the ice front no longer served as a barrier, the lakes drained, leaving a flat landscape. Other landforms have been carved into the landscape by the Mackenzie River and its tributaries. These same streams have also built up huge deltaic areas, notably at the western end of Lake Athabasca where the Peace and Athabasca rivers deposit their silt, and at the mouth of the Mackenzie River.

The Cordillera, a complex mountainous region, occupies much of northern British Columbia, Yukon, and a small portion of the Northwest Territories found west of the Mackenzie River. It is about 800 km wide and extends northwestwards to the Alaska border. The Cordillera is distinguished by its mountainous terrain, though it includes plateaus, valleys, and plains. It also contains glaciers and icefields. Its rugged mountains, particularly the Rocky Mountains which extend northwards to the Liard Basin, are in sharp contrast to the flat-lying rocks of the

Interior Plains. These mountains were formed by severe folding and faulting of sedimentary rocks. The highest mountain in Canada, Mount Logan, is part of the St Elias Mountains. It has an elevation of nearly 6,000 metres.

During the Wisconsin ice age, some 25,000 years ago, the Cordillera was severely glaciated. Alpine glaciers created arêtes, cirques, and U-shaped valleys. Many glaciers are still active, though their rate of movement is relatively slow. For most of the Cordillera, the land became ice-free some 10,000 years ago and in this postglacial period, river terraces, alluvial fans, flood plains, and deltas were formed. Periglacial landforms are commonly found in the Arctic.

The Hudson Bay Lowland is a low, flat coastal plain which has recently emerged from the sea. The surface of this lowland consists of recently deposited marine sediments combined with reworked glacial till. These deposits accumulated in the postglacial period when the lowland was under sea water (part of a much larger Hudson Bay). With the removal of the ice sheet some 7500 years ago, the process of isostatic rebound came into play. Gradually, more and more of the Hudson Bay Lowland was exposed, creating a distinct geomorphic region. Its Pre-Cambrian bedrock is masked entirely by glacial and marine sediments. The inland boundary of this geomorphic region is marked by low elevations of up to 180 metres, indicating the maximum amount of isostatic uplift for this region. Since isostatic movement is continuing at a rate of around 70 to 130 cm/100 years, the Hudson Bay Lowland is expected to increase its size at the expense of Hudson Bay. This intriguing notion is discussed more fully in Vignette 2.4.

Arctic Lands, found in much of the Arctic Archipelago, are a complex geomorphic area comprising lowlands, hilly terrain, and mountains. The cold, dry climate results in many periglacial features, which provide a distinctive appearance to the land. Rock-scattered bedrock and patterned ground are widespread in the lowlands while alpine glaciers occur in the more mountainous landscapes. Rolling to hilly terrain underlain by permafrost is affected by slumping, caused by solifluction or gelifluction: 'soil flow' down a sloping frozen surface (Trenhaile 1990: 50–1). Hilly and mountainous landforms are found in the Queen Elizabeth Islands, although lowlands occupy much of Banks and Victoria Islands. Most of Canada's glaciers are found in this region.

Glaciation

Most of the northern landforms have been glaciated. This process took place over some 25,000 years and involved both glacial erosion and deposition. Glacial erosion took place as the massive ice sheets were advancing while deposition occurred as these ice sheets melted. The Canadian Shield, for example, still shows many signs of glacial erosion.

Vignette 2.4   Hudson Bay and the Northwest Passage: The Shape of Things to Come

The Hudson Bay area, and indeed much of arctic Canada, is rising, due to glacio-isostatic uplift (recovery of the crust consequent upon the removal of the weight of the Wisconsian ice sheets). The maximum expected uplift will alter the familiar shape of Hudson Bay. Two points should be stressed: (1) even at present rates of uplift (0.7 to 1.3 metres per century), it would take at least 12,000 years for the future shape of Hudson Bay to be achieved; and (2) since isostatic uplift is expected to decrease with time, it would take very much longer for the new but smaller Hudson Bay to develop.

In a similar fashion, the Northwest Passage will be closed to shipping in the year 3,100 A.D. By then, isostatic rebound will reduce Simpson Strait between King William Island and Adelaide Peninsula to a narrow gut only 200 metres wide and barely 2 metres deep.

Adapted from Barr 1972:64; and Barr 1986.

Over the last two million years, the North has known several ice ages (Vignette 2.5). The last one, the Wisconsin, began some 25,000 years ago. During the Wisconsin ice age, two huge ice sheets, the Laurentide and the Cordillera, covered most of North America. These ice sheets reached a maximum thickness of four and two thousand metres thick respectively. In comparison, the largest ice caps in Canada are on Ellesmere Island and have a thickness of nearly one thousand metres. Some time around 18,000 years ago, as the climate warmed, these huge ice sheets began to retreat. Soon an ice-free corridor appeared along the eastern edge of the Cordillera ice sheet, connecting the unglaciated areas in Yukon with the rest of ice-free North America. By about 10,000 years ago, most of these two ice sheets had melted.

During this process of advance and retreat, two geomorphic processes took place. First, the advancing ice sheet caused glacial erosion; later, the retreating ice sheet deposited debris on the land. Glacial erosion took various forms, such as scraping off the unconsolidated material and pluck-ing out huge chunks of bedrock. Where the bedrock was highly resistant, the rock was scraped and scoured. In the mountains, alpine glaciers moved quickly down-slope and had a much greater erosional effect than did continental glaciers. Evidence of such massive erosion is found in U-shaped valleys in the Cordillera and the fjorded coastline of Labrador and Baffin Island. Glacial features of deposition consist of water-sorted deposits and unsorted deposits. As the ice sheets melted, some of the material contained in the ice was discharged into running melt water and glacial lakes. The debris held in the ice sheets was deposited on the land; in some cases, glacial melt waters sorted the debris or till into eskers and outwash plains. Eskers–long, narrow ridges of sorted sands and gravel–

were deposited from melt streams within the decaying ice sheet. Some eskers are over 100 kilometres in length. The most common glacial deposit is ground moraine. Ground moraine or till is unsorted material deposited by a melting ice sheet or glacier. Till often covers large areas while drumlins often appear as clusters of low hills, shaped by the flow of the ice. Most drumlins are believed to have formed a short distance behind the ice margin just prior to deglaciation and therefore record the final direction of ice movements (Dyke and Morris 1988: 86). As the massive ice sheets melted, enormous quantities of water were released. These melt waters either overtaxed the existing southern flowing drainage system or formed glacial lakes. Often, these glacial lakes were formed when the northward flowing rivers were still blocked from reaching the sea by the remaining ice sheet. The largest glacial lake, Lake Agassiz, occupied much of Manitoba and adjacent lands.

Glaciation has created landforms which have both positive and negative impacts on economic development. In particular, glaciation has affected road building. For example, the scraping of the Canadian Shield in northern Saskatchewan has created a northeast-southwest alignment of the Precambrian rock outcrops. Since highways from the south tend to head north or northwest, this requires building roads against the grain of the land. On a positive note, eskers and other sorted glacial deposits found in northern Saskatchewan are a much valued source of sand and gravel for highway construction.

Periglacial Features

Periglacial features are widespread in the North. They are associated with cold climates where freezing-and-thawing takes place frequently. While originally limited to landforms near the Pleistocene ice sheets and alpine glaciers, periglacial features are found over one-fifth of the earth's surface (French 1976: 3). Relic periglacial features are common to most of glaciated North America (Graf 1983). Well-established periglacial features are restricted to extremely cold areas with sufficiently low precipitation to prevent the formation of glaciers. Under these conditions, frost action is the dominant geomorphic process and various forms of patterned ground are formed. Except for areas of solid bedrock, patterned ground is widely found in the Arctic Lands geomorphic region. Patterned ground refers to symmetrical forms, usually polygons, caused by intense frost action over a long period of time. Ice-wedges are commonly associated with patterned ground (Vignette 2.6). Patterned grounds also embrace frost-sorted circular or polygon patterns of stones and pebbles. The general process of frost heave is for coarse stones to move to the surface and outwards (Sugden 1982: 103). Another even more dramatic periglacial feature, though not a common one, is the pingo. This ice-core hill is found in wet, flat areas of Canada and Alaska where permafrost is extensive.

Vignette 2.5   Contemporary Glaciers

Over the past two million years, four distinct ice ages have occurred. The latest ice age, the Wisconsin, began about 25,000 years ago. It ended some 10,000 years ago. Today, glaciers exists in two areas of the Canadian North, in the Cordillera and on the arctic islands, including Baffin Island. Large masses of ice called glaciers, icefields, and ice caps are the remnants of the late Wisconsin ice cover which consisted of the Cordillera and Laurentide ice sheets. In the Canadian North, there are around 200,000 km^2 of land covered by glaciers. Approximately 75% of the area occupied by glaciers is found in the Arctic Archipelago, 15% in Baffin Island, and 10% in Yukon.

The distinction between the types of ice masses is generally as follows: *ice caps* flow outwards in several directions and submerge most or all of the underlying land, while a *glacier* flows in one direction and is normally confined to a valley. *Icefields* refer to a number of ice caps which have coalesced. Today, they are located in the St Elias Mountains and the adjacent mountain ranges of southwest Yukon and British Columbia ; on Baffin Island (Penny and Barnes ice caps), Devon Island (Devon Ice Cap), Axel Heiberg Island (Franz Muller Ice Cap), and Ellesmere Island (Prince of Wales, Sydkapand, and Agassiz ice caps and an unnamed one). These ice caps and glaciers occupy less than 5% of the Arctic while those in the Cordillera form less than 1% of the Subarctic. The Cordillera also contains rock glaciers: glaciers have been covered by debris from rock fall and from movement of glacier debris to the ice surface. There are two types of rock glaciers: those with cores of glacial ice, and those with interstitial ice (Johnson 1988: 277). When glaciers receive a thick and complete cover of debris, their rate of melting is slowed.

Pingos often occur when an ice lens in permanently frozen ground is nourished by extraneous water. Over time, the ice lens grows and pushes itself upwards, forming a mound or hill.

Permafrost

Permafrost, perennially frozen ground, is found in almost all of the Canadian North. It is defined as ground remaining at or below the freezing point for at least two years. At particular sites, the depth of permafrost may exceed several hundred metres while at more southerly sites, its depth may be less than 10 metres. The greatest thicknesses in Canada are over 1000 metres in areas of Baffin and Ellesmere islands. As well, pockets of permanently frozen ground may lie well below the earth's surface (Figure 2.7).

Permafrost has existed for thousands of years in the Canadian North and, unless the vegetation cover is disturbed, it remains unaffected by the summer melting of the surface of the ground. The summer melting of the upper layer of ground (called active layer) is a normal occurrence. The

Figure 2.7
Permafrost Zones of the Canadian North

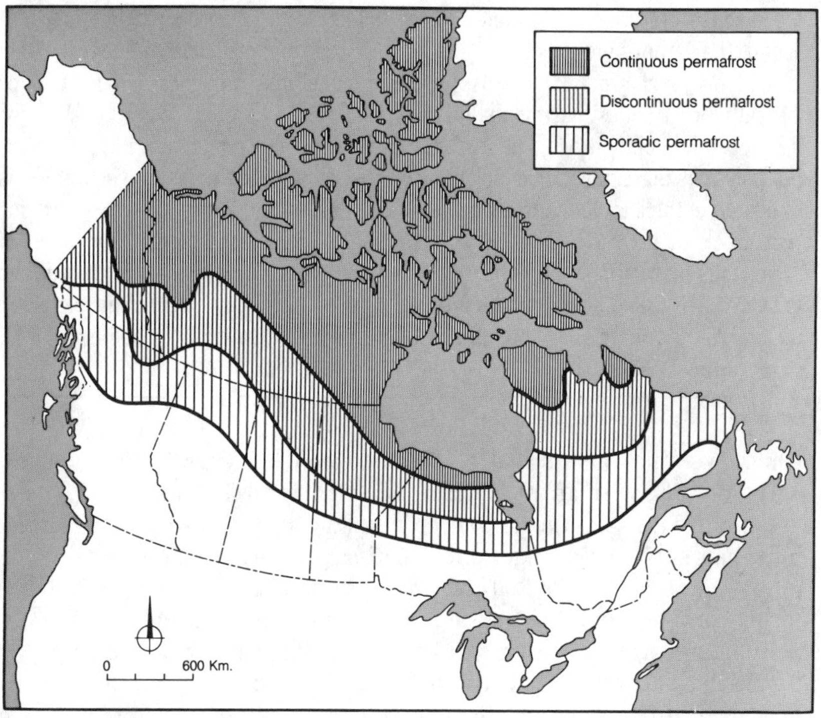

SOURCE: Adapted from Trenhaile 1990: 122.

thickness of the active layer varies from a few centimetres in the Arctic to over 10 metres in the Subarctic. The southern extent of permafrost is associated with the mean annual air temperature isotherm of 0°C (Williams 1986: 3). Although permafrost is theoretically present near this isotherm, there is a transition zone consisting of a mixture of frost-free and frozen ground between 0°C and −5°C annual isotherms.

Since permafrost varies in its depth and geographic extent, it is divided into four types–continuous, discontinuous, sporadic, and alpine permafrost. An area is classified as having continuous permafrost when over 80% of its ground is permanently frozen; for discontinuous permafrost, 30 to 80% must be frozen; and for sporadic permafrost, less than 30%. Alpine permafrost is not defined by the percentage of permanently frozen ground but by its presence in a mountainous setting. Such permafrost is found in British Columbia and Alberta south of the zone of sporadic permafrost (Figure 2.7). In terms of geographic distribution, continuous permafrost is found where the mean annual temperature is around −7°C

Vignette 2.6   Ice-Wedge Cracks

Ice wedges are widely distributed in the colder parts of permafrost areas of the world. The wedges are typically V-shaped in section and the ice is vertically foliated. Large polygonal ground patterns usually indicate the presence of ice wedges beneath the troughs delimiting the polygons. The polygonal patterns are best developed on poorly drained peaty flats. Ice wedges also occur on many hill slopes but polygons are usually inconspicuous or absent, because mass movement tends to obliterate the ice-wedge troughs. In the western Arctic, active ice wedges (those that crack at least once every 5 to 10 years) are most abundant in areas of continuous permafrost where the mean annual ground temperature near the surface is below −5°C and snow depths are less than 50 cm. They are best developed in unconsolidated material, although some ice wedges also grow in bedrock.

SOURCE: Mackay 1989: 365

while discontinuous permafrost lies between −5°C and −7°C (Figure 1.2). Sporadic permafrost is often found between 0°C and −5°C.

Permafrost affects the land surface by slowing the growth of vegetation, impeding surface drainage, and creating periglacial landforms such as pingos and thermokarst features. While permafrost and periglacial landforms are often found in the same geographic area, permafrost refers to permanently frozen ground while periglacial features are created by freeze-thaw action. Permafrost, a natural feature of the North, has only become a 'problem' since efforts have been made to develop the North by building roads, pipelines, and towns. The design of these human-made features must take into account the presence of permafrost. Exposure of ice-rich ground during construction can result in retrogressive thaw slumps (Burns and Lewkowicz 1990) and such slumps can prove costly. Permafrost has made northern construction 'a matter of geotechnical science and engineering' (Williams 1986: 27).

Northern Seas

The seas of the Canadian North extend from the Beaufort Sea to the Labrador Sea and include the waters of Hudson Bay. These water bodies have less impact on the northern hydrological cycle than might be expected. The reason for their limited impact is that most are covered by ice for more than half the year; waters lying in much of the Arctic Archipelago have permanent ice cover. Polar pack ice (ice which has not melted for at least two years) covers most of the Arctic Ocean. Pack ice is much thicker than other ice cover, making navigation by ship most difficult.

On the Canadian side of the Arctic Ocean, the circulation is clockwise. The current carries pack ice and occasionally ice islands. Cold surface

Table 2.2
The Drainage Basins of Canada and their Streamflows

Drainage Basin	Area (million km²)	Streamflow (m³/s)
Hudson Bay	3.8	30,594
Arctic	3.6	20,491
Atlantic	1.6	21,890
Pacific	1.0	24,951
Total	10.0	105,135

SOURCE: *Canada Year Book 1988*: 1–7; and Laycock 1987: 32.

water flows from the Arctic Basin into the Atlantic Ocean around Baffin Bay where it mixes with the warmer Atlantic waters, forming a subarctic water. Baffin Bay is connected to the Arctic Ocean through Nares Strait, Jones, and Lancaster sounds and to the Labrador Sea through Davis Strait. Icebergs, formed by calving off Greenland glaciers, often are carried by the Labrador Current to Newfoundland. During winter when there is extensive ice cover, a polynya or open-water area (called 'North Water') exists in the northern part of Baffin Bay. The explanation for this open water still eludes physical scientists.

Northern rivers, such as the La Grande, Mackenzie, and Churchill rivers, empty vast quantities of fresh water into these seas. One impact of this fresh water is to stratify the ocean waters, that is, river water tends to overlie the colder ocean waters. This stratification of ocean waters results in surface layers having a lower salt content. Another effect of these fresh waters is to provide a suitable estuarian habitat for marine life such as bowhead whales.

Northern Hydrology

The northern hydrological cycle is strongly affected by the cold environment. For most of the year, precipitation is slight and northern waters are frozen. During the short summer period, however, the hydrological cycle becomes more active. Added to the release of water through melting ice and snow, most precipitation falls in the summer period.

Though annual precipitation is low, water resources in the North make up the bulk of Canada's water reserves. These resources are found in four drainage basins–the Arctic, Hudson Bay, Atlantic, and Pacific (Figure 2.8). The river systems found in each drainage basin empty their waters into the three oceans surrounding Canada. The Arctic and Hudson Bay drainage basins form nearly 75% of the area of Canada and, despite relatively low precipitation over much of this area, these two basins account for almost 50% of the streamflow (Table 2.2).

The hydrological cycle peaks during the spring runoff. Nearly one-third

Figure 2.8
River Drainage Basins of Canada

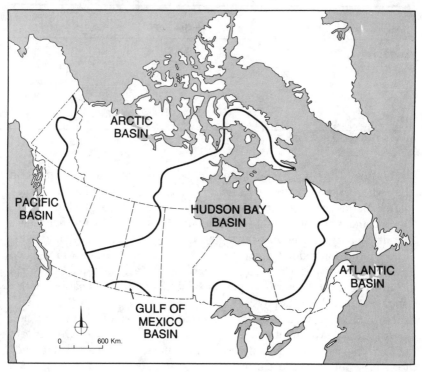

SOURCE: *Canada Year Book 1988*: 1–7.

of the annual precipitation falls as snow and is released during the spring melt. As well, fresh-water bodies, frozen in the winter, release their water in the spring. In terms of runoff volume, the Mackenzie River is the most important river in the North. Since a large proportion of its headwaters are located in British Columbia, Alberta, and Saskatchewan, spring melting occurs south of 60°N. while the lower reaches of the Mackenzie River are still frozen. Ice jams frequently occur, causing widespread flooding.

To sum up: the physical geography of the North is dominated by a cold environment. This environment embraces climatic, vegetation, and soil conditions; permafrost, periglacial features, and glaciers. Since the last ice age, the Arctic and Subarctic biomes have been shaped by the polar climate, resulting in continental-size zones of similar natural vegetation, soils, and biological life. These two environments also represent cultural regions. Over time, the Indian and Inuit people developed cultures and technologies suitable for living as hunters and gatherers in each region. More recently, each environment poses a different set of physical challenges to resource developers. Much of this challenge is related to the fragile nature of the North, particularly the Arctic. The slow rate of natural growth means that polar lands and seas take much longer to recover from industrial accidents, such as an oil spill, than do more temperate places.

Selected Readings

Heginbottom, J.A. (co-ordinator), 1989. 'A Survey of Geomorphic Processes in Canada', in *Quaternary Geology of Canada and Greenland*, ed. R.J. Fulton. Ottawa: Minister of Supply and Services.

Trenhaile, Alan S., 1990. *The Geomorphology of Canada*. Toronto: Oxford University Press.

Williams, Peter J., 1986. *Pipelines & Permafrost: Science in a Cold Climate*. Ottawa: Carleton University Press.

Young, Steven B., 1989. *To the Arctic: An Introduction to the Far Northern World*. New York: Wiley.

Chapter 3

THE HISTORICAL BACKGROUND

The first people to set foot on North America were Old World hunters who crossed the Beringia land bridge into Alaska (Vignette 3.1). But just when these Palaeo-Arctic hunters reached Alaska and later the interior of North America remains in doubt. Archaeological evidence confirms that Palaeo-Indians occupied southwestern United States some 11,000 years ago, suggesting that their ancestors migrated along the ice-free corridor between the Cordillera and Laurentide ice sheets some 12,000 years ago (Vignette 3.2). Other archaeological findings such as those in Yukon near Old Crow suggest that humans may have been present much earlier, perhaps before the last ice age began, If so, these early peoples could have migrated southwards and occupied the Americas some 30,000 years ago.

Occupying Northern Canada

The Palaeo-Arctic hunters quickly occupied lands south of the ice sheets. Their main source of food remained the woolly mammoth and other big game animals. Around 10,000 years ago, the woolly mammoth became extinct and its hunters had to adjust to new circumstances. Unfortunately, the archaeological record is very sketchy and no clear picture can be presented. Suffice to say that the Indian tribes found in North America by Europeans either evolved from these early Asian migrants or from

Vignette 3.1 Beringia

During part of the last ice age, a land bridge joined Asia to North America. This land is called Beringia. Much of it is now some 100 metres below sea level. Around 18,000 years ago, the ice sheets reached their maximum extent. At that time, huge quantities of water were locked into the ice sheets, causing the sea level to drop by as much as 150 metres, and in the process exposing the Beringian land bridge. Later, as the climate began to warm, two major physical events took place. The ocean waters rose, covering the land bridge once again, and an ice-free corridor appeared between the Laurentide Ice Sheet and the Cordillera Ice Sheet, the coalesced alpine glaciers flowing out of the Rocky Mountains.

Vignette 3.2 The Clovis Theory

According to the Clovis theory, Palaeo-Arctic Man entered North America (Alaska and Yukon) around 12,000 years ago. At first, lands to the south were blocked by a massive wall of ice but soon Laurentian and Cordillera ice sheets began to melt. An ice-free corridor was formed, allowing game and man to move southward. Armed with a spear which had a distinctive fluted projectile point, these Palaeo-Indians quickly spread southwards in search of game. Archaeological evidence clearly demonstrates that Palaeo-Indians were hunting the woolly mammoth and other big game animals in the Great Plains and the Southwest of the United States as early as 11,000 years ago. Within 2,000 years, descendants of these primitive hunters had reached the southern tip of South America. The empirical support for this theory of Early Man's movement into North America is based on archaeological evidence, particularly the stone points gathered at the famous mammoth-kill sites near Clovis, New Mexico and similar finds in other parts of North America.

Adapted from Estabrook 1982: 87–8.

more recent migrants. During the last 6,000 years northern Canada was free of ice and the Subarctic became the home of Indian tribes.

Roughly a thousand years later, a second wave of Palaeo-Arctic hunters, the Palaeo-Eskimos, crossed the Bering Strait from Siberia to Alaska. By this time, ice sheets had retreated from the arctic coast, allowing the sea-based hunting economy of the Palaeo-Eskimos to flourish. These early migrants are named Denbigh after an important archaeological site in Alaska. Their culture, the Arctic Small Tool Tradition, was based on the use of flint to shape bone and ivory into harpoons and other tools. As sea hunters, they occupied arctic coastal lowlands and eventually reached the coast of Labrador and Newfoundland. Over time, these people were replaced by a new wave of Palaeo-Eskimo migrants–the Dorset peoples–who in turn were replaced by the Thule peoples. While the archaeological story is incomplete, two major cultural changes have been identified. The first one saw the Denbigh culture evolve into, or replaced by, the Dorset culture. The Dorset people lived in semi-permanent houses built of snow and turf, heating them with soapstone oil lamps.

Around 1000 A.D, the Thule culture appeared. The Thule people originated in Alaska and gradually spread across arctic Canada, replacing the Dorset inhabitants. In the open waters around the arctic coast of Alaska, the Thule hunters had developed the technology and skills necessary to hunt the large bowhead whales. Their migration eastward may have been caused by population pressures or by the increase in open water in the Arctic Sea caused by a warming of the arctic climate. In any case, the Thule peoples quickly spread across the Canadian Arctic using their innovative transportation system of skin boats (kayak and umiak) and dog sleds to harvest seal, walrus, caribou, and the bowhead whale.

After 1200 A.D., the climate began to cool, making ice cover more extensive. The climate continued to cool and the period between 1450 and 1850 is referred to as 'the Little Ice Age'. Ice cover became very extensive and bowhead whales ceased to appear in parts of the Arctic. These new circumstances required a shift in Thule hunting strategy; no longer could they remain in one place for long periods of time. Now they became more mobile and formed smaller groups to hunt the seal and caribou. By around 1700, the Thule whaling economy seems to have disappeared (Crowe, 1974: 18). If this analysis is correct, then the Vikings and early European explorers like Martin Frobisher probably encountered Thule people.

Aboriginal Peoples at the Time of Contact

When Columbus 'discovered' America, Canada had been occupied by Aboriginal peoples for thousands of years. In the North, the hunting cultures had become finely tuned to their environment. The general distribution of Native people at pre-contact is shown in Figure 3.1. At that time, the Algonkian Indians lived in the eastern Subarctic while the Athapaskan Indians occupied the western Subarctic. To their north lived the Thule peoples who eventually became the modern-day Inuit.

The geographic location of individual tribes is much more problematic, but these hunters likely had occupied the territory shown in Figure 3.1 for some time, perhaps centuries. The geographic stability of the pre-contact hunting societies can only be guessed at because archaeological evidence is so limited. Only in the Arctic is there evidence of the replacement of one group by another.

Prior to contact with Europeans, Indians and Inuit living in northern Canada engaged in a variety of nomadic hunting economies, securing food by hunting, fishing, and gathering. While dried fish and meat could be stored, in general members of hunting societies were constantly seeking food. The main source of food was big game. For many, caribou was the

Figure 3.1
Areas Occupied by Aboriginal Peoples at Contact

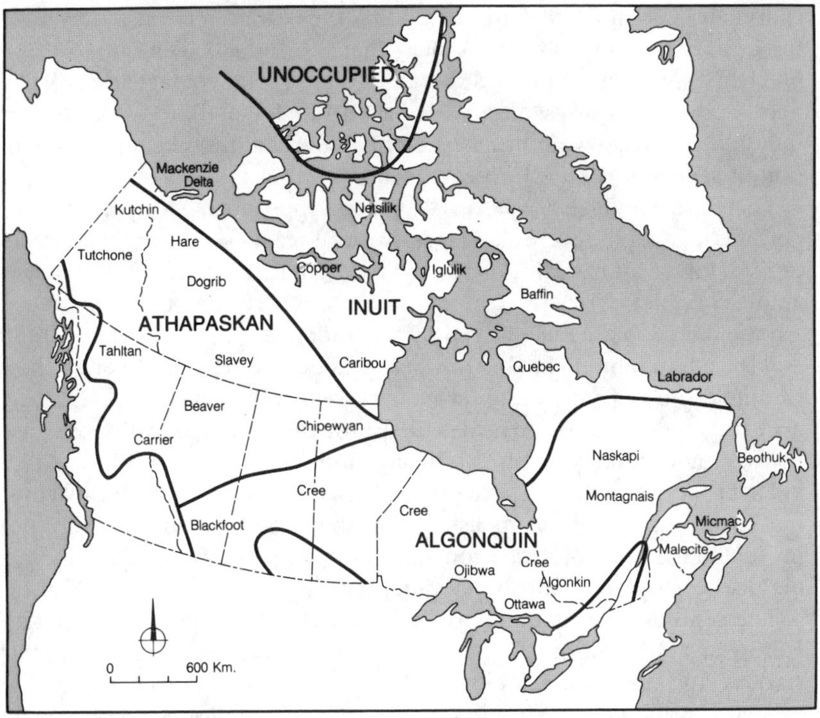

SOURCE: Harris 1987: plate 11, and McMillan 1988: xii.

preferred food; whale blubber and seal meat were popular with the Thule Eskimos. These hunting peoples moved according to the seasonal migrations of wild animals. Such a migratory hunting strategy is a form of sustainable resource use. The necessity of moving from one area to another minimized the depletion of local resources, and cultural taboos controlled over-hunting or fishing.

The hunting economy had many attractions but it also contained an element of risk. For example, an unforeseen shift in the annual migration route of caribou could expose hunters and their dependents to severe shortages of food and perhaps starvation. As Ray (1984: 1) observed, 'opinions are divided whether hunting, fishing, and gathering societies generally faced a problem of chronic starvation–the more traditional viewpoint or whether they were the original affluent societies.' Krech (1978: 101) examined population change among the Kutchin and concluded that 'Whether starvation was a significant cause of recurrent mortality among precontact Kutchin and other northern Athapaskans, as some would argue, is problematic.' Since oral histories and archaeological evidence do not answer the question of 'food shortages' in the pre-contact period, one can only speculate on the severity and frequency of such shortages. Morantz (1984: 63) noted that the written records left by early fur traders and others made it clear that the Indians living in northern Quebec were troubled by food shortages shortly after contact and possibly before contact: 'Even with a varied diet, the James Bay people of two hundred years ago were not assured of a consistent food supply. Starvation, it seems, was known to each generation.'

There are a number of reasons why food supply might vary periodically, perhaps leading to severe shortages. Natural reasons include the destruction of forest and wildlife by forest fires, and low cycles in the small-animal population. After contact, much more serious problems faced Aboriginal peoples. Encouraged by fur traders and the promise of trade goods, Indian hunters broke with past practices and over-exploited their wildlife resources. Game and fur-bearing animals became scarce, creating a need for new lands. Armed with European weapons, tribal warfare became more bloody as fur-producing lands were invaded. But the real killer in the post-contact period was disease brought to the New World by Europeans. Smallpox, measles and other 'foreign' illnesses were not present in the pre-contact period and all too frequently these communicable diseases devastated tribal populations.

The general location of Aboriginal peoples at the time of contact with Europeans is based on the diaries and accounts of early explorers, fur traders, and missionaries. How long these territorial arrangements had existed is not known. What is known is that the very process of contact with Europeans triggered a series of population shifts. Generally, those aboriginal tribes that obtained weapons from the European traders began to dominate their neighbours, forcing them to flee from their traditional lands. In other cases, tribes, such as the Beothuks, the Mackenzie Inuit, the Sadlermiut, and the Yellowknifes, disappeared and their lands were occupied by Europeans or by other tribes.

Until recently, the number of Aboriginal peoples was, at best, a rough approximation. From the time of contact to the late nineteenth century, these estimates were based on accounts of explorers, fur traders, and government officials. One estimate by Heidenreich and Galois (Harris 1987: Plate 69) suggests that, in the early seventeenth century, there were about 250,000 Native people living in what is now known as Canada. How many lived in northern Canada is unknown. One estimate suggests that some 50,000 Indians and Thule Eskimos occupied the Subarctic and Arctic areas of Canada at the time of contact (Crowe 1974: 20). By the early 1820s, the Native population in Canada was estimated at 175,000, indicating a substantial decline from the earlier figure (Harris 1987: Plate 69).

Within the Arctic, there was considerable diversity among the Eskimos. In the 1500s, for example, there were some 10,000 people living in the Arctic. These were divided into ten distinct Thule or Eskimo tribes–the

Labrador, Ungava, Baffin, Polar, Sadlermiut, Iglulik, Caribou, Netsilik, Copper, and Mackenzie. Each group may have contained between 500 and 1000 members who were spread out over a vast but identifiable territory. During this time, small extended family units lived on the land. For short periods, usually during winter sealing, these families would come together. This geographic pattern of life was dictated by the varying capacity of the land to provide opportunities to hunt successfully for food.

Today, seven of these Inuit tribes still exist in Canada. European diseases may have caused the demise of the Sadlermiut and the Mackenzie tribes, while the Polar Eskimos now reside in Greenland. More is known about the Mackenzie Inuit and historical records indicate that smallpox and influenza brought to their area by whalers decimated their numbers. The present Inuit living along the arctic coast near the mouth of the Mackenzie River were originally from Alaska and their parents migrated into Canada after the lands became vacant.

In the Subarctic, there were some 40,000 Indian peoples at the time of contact. They lived by hunting, fishing, snaring, and gathering wild plants and berries. Men did the big game hunting while women snared hare, fished, prepared meals, and made clothes from hides. Like the Inuit, the Dene occupied a vast territory, moving from place to place in search of game. Such movements were in response to seasonal shifts in animal populations, particularly those of the caribou. Their hunting lifestyle demanded mobility and this limited material possessions. On the whole, these bands did not have formal chiefs before European contact, though people aligned themselves with persons who manifested leadership abilities.

There were at least 22 distinct tribes in northern Canada when the French colonies were established along the St Lawrence. From a linguistic perspective, all but two of these tribes fall into three main groups–two Indian languages in the Subarctic and one Inuit language in the Arctic. In the Subarctic, one group speaks Athapaskan languages and the other group Algonkian languages. Each linguistic group was found in a particular geographic area at the time of European contact, the Algonkian-speaking in the eastern portion of the Subarctic and the Athapaskan-speaking in the western part of the northern forest. This broad spatial pattern has changed little over time, though individual tribes have relocated. The Cree, for instance, spread into Manitoba, Saskatchewan, and Alberta. The eight northern Indian tribes speaking Algonkian languages include the Cree, Micmac, Montagnais, Naskapi, and Ojibwa. The Athapaskan linguistic family consists of 15 distinct spoken languages in Canada. These are the Beaver, Carrier, Chilcotin, Chipewyan, Dogrib, Han, Hare, Kaska, Kutchin, Sarcee, Sekani, Slave, Tagish, Tahltan, and Tutchone.

The Arrival of Europeans

The first Europeans to land on North American soil were the Vikings. During the tenth century, the Vikings regularly travelled from Greenland to Ellesmere and Baffin islands to trade with either the Dorset or Thule peoples. They also attempted to establish settlements along the coast of Labrador and Newfoundland. The first authentic Norse site found is located at L'Anse aux Meadows on the northern tip of Newfoundland's Great Northern Peninsula.

Some five hundred years later, the Viking discoveries were all but forgotten, leaving Europe's geographic knowledge of the New World scanty at best. Explorers seeking not North America but the rich spice lands of Cathay brought word of a new land. Reports of the new land's great riches, based on information obtained from local Indians, may have been misinterpreted or embellished in hopes of securing financial support for another voyage. In 1536, Jacques Cartier returned to France with a report of a golden 'Kingdom of Saguenay' (Trudel 1988: 368). This report was based on information supplied by Indians and supported by the Iroquois chief Donnacona whom Cartier had captured and took back to France. Cartier thought that he had discovered diamonds and gold from this mythical kingdom. These riches, upon careful examination in France, turned out to be quartz and iron pyrites–'fool's gold'.

In a similar vein, Martin Frobisher set sail from England to find the Northwest Passage to the Orient. Instead, he discovered Baffin Island. His two sea voyages in 1576 and 1577, mark the beginning of British exploration of arctic Canada. As with most early voyages, contact with the local peoples was established and trade often followed. However, contact between these two very different peoples did not always end on a friendly note. For instance, after landing on the shore of Baffin Island in 1576, five of Frobisher's crew were captured by local Inuit during a skirmish and were never seen again (Cooke and Holland 1978: 22–23). The following year, Frobisher lured an Inuk in his kayak to come near his ship. Captured and taken to England, the Inuk's skills at handling his kayak while hunting the royal swans were observed by the Queen (Crowe 1974: 65). After a few years, he died of natural causes in England.

During the remainder of the sixteenth century and in the early seventeenth century, many explorers including Davis, Hudson, and Button continued to search for a route to the Orient. In the course of this search they explored the waters of Baffin Bay and Hudson Bay. These explorers failed to find a route to the Orient but the 'casual' trade with Indians for furs marked the beginnings of a highly profitable enterprise and to the formation of an English fur-trading company–The Hudson's Bay Company.

The Fur Trade

For over 300 years, the fur trade was the most powerful economic, social, and political force in the Canadian North. It exerted a firm grip on early Canadian history and, through the search for fur, outlined the geography of Canada. By 1760 the fur trade was the dominant economic activity in New France and Rupert's Land (Figure 3.2). The natural highways were the rivers while the forested lands of the Canadian Subarctic housed beaver and other fur-bearing animals. Canoes and, later, York boats were the vehicles of the fur trade. Trading posts were located at strategic points on these water routes (Vignette 3.4).

The fur trade dictated the nature of relations between Natives and Euro-Canadians. As the fur trade evolved, so did this relationship. At first each party needed the services of the other; they were more or less equal. With time, however, Indians became more and more dependent on the fur economy. With the industry's decline their dependency shifted to the federal and provincial governments.

Vignette 3.3   Beaver Hats

For several centuries, the fur trade was based on the beaver skin, highly valued in Europe as a material for hats. This demand, driven by fashion, was both the strength and weakness of the fur trade. Prices of beaver and other furs were subject to sharp fluctuations and these changes could greatly alter the number of furs required for European goods.

The beaver was sought not for its pelt but for fur-wool, the layer of soft, curly hair growing next to the skin, which had to be separated from the pelt and from the layer of longer and stiffer guard hairs. This fur-wool was then felted for cloth or hats. The use of beaver fur-wool for hats became especially important. In England, for example, Spanish and Dutch immigrants popularized the habit of wearing hats in place of woollen caps early in the sixteenth century. Thereafter, no amount of sumptuary legislation could stem the decline of cap making. Wearing caps became a hallmark of the lower classes. For those of higher status, shape and type of hat became a barometer of political allegiances. The Stuarts and their followers favoured the high-crowned, broad-brimmed, and squarish 'Spanish beaver'. Only in the early nineteenth century did the beaver hat go out of fashion, in favour of hats made of silk and other materials.

SOURCE: Wolfe 1982: 159–60.

The story of the fur trade in Canada began with the Portuguese fishermen who traded with Indians along the shores of Newfoundland, but its greatest impact was on New France. As early as the sixteenth century, French fur traders located along the St Lawrence River recognized that the best furs came from the Subarctic where the long cold winter resulted in long-haired fur pelts. Geography provided easy access to these northern Indians by means of southward-flowing rivers, such as the St Maurice. Late in the next century, the British established a series of trading posts on the shore of Hudson and James bays. For over a century, these two nations competed for the highly prized furs of the northern Indians. Even when French Canada fell to the British in 1763, two centres, London and

Figure 3.2
European Spheres of Influence, *circa* 1760

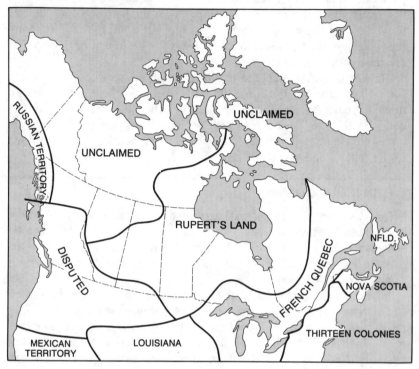

Adapted from Watson 1963: 171 and Harris 1987: plate 44.

Montreal, continued to compete for northern furs. Montreal fur traders, now a blending of French Canadian and Scottish traders, continued to use their strategic geographical location and connecting river systems to intercept Indians on their way to the Hudson's Bay trading posts at York Factory and other trading posts on the west coast of Hudson Bay.

The nature of this fierce competition forced the fur traders to move further inland, building trading posts in western Canada and the territories in hopes of attracting local Indians to the posts. Consequently, the fur traders established a network of trading posts across the Subarctic and, in doing so, firmly involved local Indians in the fur trade economy. The inland posts also tied Indian tribes to the areas around fur trading posts, thereby reducing the range of their hunting activities. This smaller hunting area was often over-exploited, causing the Indians in times of shortage to seek food from their fur trader.

The fur trade, by introducing European goods, institutions, and values to the Aboriginal peoples, caused many changes to the traditional native way of life and rearranged the geographic distribution of many Indian tribes. Traditional native material culture was greatly diminished by substituting European goods for those produced from the local environment. Indian women found an iron needle much easier to sew skins than a bone one. Similarly, iron kettles and copper cooking pots replaced the traditional wooden and bone ones, making domestic life easier. European fire arms, iron knives, and axes were in great demand, replacing traditional weapons such as bows and arrows. Such a weaponry advance often tipped the balance of power between those Indian tribes with muskets and those with bow and arrows. The fur trade had implications, therefore, for tribal relations, control of territory, and access to fur trading posts.

In a similar fashion, the native spiritual world was challenged by Europeans. Most spiritual changes occurred when missionaries brought Christian beliefs into the North. These missionaries accelerated the process of cultural change. By the middle of the nineteenth century, Anglican and Roman Catholic missionaries had established themselves throughout the North. Usually, Anglican and Roman Catholic priests located their missions adjacent to the fur posts. Their objective was simple–to convert the Indian peoples to their version of Christianity. It was not long before the shamans/medicine men no longer held a central position in Native society. In the nineteenth century the introduction of boarding-school 'European-style' education by the two churches was an important step toward assimilation and the eroding of traditional Native beliefs and languages. Until the post-World War II years, the federal government encouraged the Anglican and Roman Catholic missions to educate native children. The languages of instruction were English and French and the children were not allowed to speak their own language. Regardless of the 'good intentions' of the missionaries, their efforts further weakened the confidence of Aboriginal peoples in their own culture, causing them to be more dependent on but at the same time not part of western culture.

The fur trade drew the Indians and, much later, the Inuit into a new pattern of social and economic relations. Yet the degree of involvement of Native people varied over time and space. While the exchange of furs for trade goods provided the Indians and Inuit with access to European technology, the main impact was on those Native people involved in direct trading with Europeans. Others, more distant from the fur traders, were only marginally involved. Relations between trappers and traders were not constant over time and whenever trapping and/or hunting was unable to satisfy the needs of Natives, they turned for help from their fur trader. In this way, a rather balanced, mutually advantageous relation in which each partner was more or less equal and certainly in need of the other changed into an unbalanced dependency with Natives trapped into bartering for western tools and weapons now essential for hunting, fishing, and trapping. Goods were also needed to run the household. This dependency has culminated in a twentieth century Native welfare society which Ray (1984) argues was not a sudden event associated with the recent decline of the fur trade but rather one that has its roots in the early fur trade and the role of the Hudson Bay Company: Ray expresses the basic elements of his case:

> the Hudson's Bay Company was partly responsible for limiting the ability of Indians to adjust to the new economic circumstances at the beginning of this century. Debt-ridden, repeatedly blocked from alternative opportunities for over a century, and accustomed to various forms of relief for over two centuries, Indians became so evidently demoralized in this century, but the groundwork for this was laid in the more distant past. (Ray 1984: 17)

A map of the involvement of Native peoples in the fur trade and their resultant economies around 1820 has been created by Heidenreich and Galois. They classify the Indian economies existing at the time into seven types, ranging from 'traditional economies severely disrupted through resource depletion and the restriction of seasonal movement due to European settlement' to 'traditional economies outside European contract' (Harris 1987: Plate 69).

The fur trade brought new technology to Native Americans but it also drew them into the European and later a global economic system. With the transfer of Hudson's Bay Company lands, in 1870, the more fertile lands of Western Canada began to be settled by Europeans, displacing the original inhabitants. Indians in the Subarctic did not have to face wave after wave of agrarian settlers, but they were placed in a 'double bind' for they were dependent on fur traders who brought trade goods, and yet fur trapping and increased hunting were undermining their resource base.

The Hudson's Bay Company

In 1665, Medard des Groseilliers and Pierre Radisson convinced Prince Rupert that a great deal of money could be made by trading furs with Indians along the Hudson Bay coast. These northern Indians currently traded their furs with the French, who were based on the St Lawrence. Groseilliers and Radisson knew that ships anchored near important northern rivers would soon attract local Indians and trade would ensue. Rupert persuaded his cousin Charles II of England, and some merchants and nobles to back the venture. In 1670, the King granted wide powers to the 'Governor and Company of Adventurers' of the Hudson's Bay Company. These powers included exclusive trading rights to all the lands whose rivers drain into the Hudson Bay. These lands were named Rupert's Land.

At first, the company sent ships to trade with the Indians who gathered at the mouths of rivers draining into Hudson Bay. Once the trade was completed, the ships returned to England. The Hudson's Bay Company soon established permanent trading posts, encouraging Indians living in the interior of the country to trade there. By locating at the mouths of rivers draining Western Canada, the Bay extended its control over lands far into the interior, thereby increasing the number of Indians involved in trapping and, of course, maximizing the number of furs received. The Hudson's Bay Company was a great commercial success from the start. It was also a political success for its Bay forts provided an important foothold for the British.

The harsh climate around Hudson Bay prevented the fur traders from supplementing their food supply with agricultural products. They turned to local Indians to supply them with game and with other useful Indian products such as snowshoes and canoes. These local Indians became known as the Homeguard. Indian Middlemen became traders who exchanged European goods for furs with inland Indians. This arrangement provided great profits to the Middlemen and allowed them to control the inland Indians. Often the Middlemen Indians would not even permit other Indians to travel to the coastal trading posts.

The creation of trading posts along the coast of Hudson Bay all followed the same pattern–at the estuaries of the major rivers draining the interior of the forested lands of what is now Manitoba, Saskatchewan, Quebec, and Ontario. The most important of these trading posts was York Factory. Situated at the mouth of the Hayes River, it offered the easiest river access to the rich fur lands of the interior. This post, because of its access to the river network leading into the interior of the Subarctic, soon began drawing furs from Indians as far distant as the Mackenzie River Basin (Zaslow 1984: 6). This statement by Zaslow suggests that, by the early eighteenth century, the Hudson's Bay Company's fur hinterland extended into the 'unclaimed' territories indicated on Figure 3.2. A similar case may be made for the Russian fur trade zone because Indians in the Yukon Basin may have made their way to the Pacific Coast trading posts.

The Hudson's Bay Company's hold on the fur trade in the western interior remained firm until French traders established inland posts on the Saskatchewan River. By 1750, French competition had caused a sharp decline in the number of pelts reaching York Factory. After the fall of New France in 1763, Montreal became the strategic headquarters for a number of fur traders, first known as the 'Pedlars' and, in 1784, as the North West Company. These Montreal fur traders intercepted Indians on their way to trade furs at the Hudson's Bay Company posts on the shores of Hudson Bay. The strategy of the Montreal traders was simple–reduce the distance Indian fur trappers had to travel to reach a trading post. At first, this scheme simply called for trading with Indians hundreds of miles inland from the Bay trading posts. Later, it was necessary to establish a series of inland trading posts. One of the most important trading posts of the North West Company was Fort Chipewyan (see Vignette 3.4).

Vignette 3.4   Fort Chipewyan and Its Geographic Advantages

In 1788, the North West Trading Company established a fort on the shore of Lake Athabasca. It was named Fort Chipewyan after the Indians who hunted in the area. Fort Chipewyan was an ideal place for the furthermost post on a fur-trading route. Not only was the fort located in prime beaver country where the pelts were of the highest quality, but its location allowed both the eastern and western fur brigades sufficient time to get back to Montreal and Fort Chipewyan respectively before the rivers froze. The geographic advantage of Fort Chipewyan's location was its ready access to the Athabasca River which breaks up in the spring a month before the lake. This gave the western fur brigade an extra month for their canoe trips to rendez-vous with their eastern counterparts.

SOURCE: Donaldson-Yarmey 1989: 10.

The strategy of permanent inland posts had one serious problem, namely, provisioning. The North West Company solved the problem by supplying its inland posts with pemmican, a food made by Plains Indians. Pemmican was light and nutritious and, most importantly, would not spoil. It consisted of dried meat (usually buffalo) pounded into a coarse powder and then mixed with melted fat and possibly dried saskatoon berries. In this way, the fur trade of the late eighteenth century became dependent on food supplies from another natural region, the grasslands of the Canadian Prairies. This arrangement indirectly involved the buffalo-hunting Plains Indians and Métis in the fur trade.

The success of the Nor'Westers' inland trading posts forced the Hudson's Bay Company to meet the competition by moving inland. In 1774, the English company established its first inland post at Cumberland House on the Saskatchewan River. This action caused the Montreal traders to move further west and, in 1776, Primeau established a post on the Churchill River near the present-day settlement of Ile-à-la-Crosse (Cooke and Holland 1978: 95). In 1787, the remaining individual traders either joined forces with the now powerful North West Company or left the country. Competition between the North West Company and the Hudson's Bay Company intensified, leading to the strategy of 'matching' trading posts, i.e., if one company established a post in a new area, the other company built one nearby.

For the Indians, there were several consequences of this competition. First of all, the new post allowed inland Indians direct access to traders. Secondly, it weakened the power base of the Cree who lost their role as middlemen. The role of middlemen passed to the Chipewyans who traded with the northern Indians, namely, the Yellowknife and Dogrib Indians. Lastly, it drew local Indians more fully into the fur trade and so made them more dependent on trade goods.

The next phase in the fur trade saw a return to a monopolistic situation.

With profits down and frequent bitter struggles between rival fur traders, the British Colonial Office wanted the North West Company and the Hudson's Bay Company to settle their feuding. In 1821, a parliamentary act of the British government attempted to placate both parties by devising a coalition. The British government authorized that the name, charter, and privileges of the Hudson's Bay Company would be assigned to the new firm but that its fur traders would include those of both the North West Company and the Hudson's Bay Company.

Under the direction of George Simpson, the 'new' Hudson's Bay Company began a rationalization of its trading posts. Posts were abandoned, staff reduced, and prices increased. Soon profits rose, reaching undreamed-of levels. The Company controlled not only Rupert's Land but also the North-Western Territory. Trade again flowed through its forts on the shores of Hudson Bay.

While his main objective was to increase the profits of the company, Simpson realized that such profits could only be achieved if the Indian way of life also prospered. He supported the concept of supplying Indians with food when country food–fish and game–was in short supply. Such a policy led the Indians into an ever-increasing and now systematized state of dependency.

A new challenge to the Bay route appeared by the late 1850s when an American railway company extended rail service to St Paul on the Mississippi River. In 1859, the Mississippi steamboat, the *Anson Northrup*, arrived at Fort Garry. The route of fur shipments to England from the North changed, swinging in favour of a southern route through the United States by steamboat and rail to New York City. The southern route was even more attractive for the shipping of trade goods. By 1872, the English River (now named the Churchill River) and Athabaska brigades abandoned the Hayes River route to York Factory. Furs now went to Fort Garry (Winnipeg) and then by steamboat southward to American rail and New York. Later, when the Canadian Pacific Railway was completed, the furs were shipped along this Canadian railway.

Until the beginning of the twentieth century, fur trading was confined to the Subarctic. When European fur buyers decided that the arctic fox pelt had commercial value, the Hudson's Bay Company quickly spread their operations into the Arctic, and soon had a series of Arctic trading posts, including Wolstenholme (1909) on the Ungava coast, Chesterfield (1912) on the Hudson Bay coast, Aklavik (1912) near the mouth of the Mackenzie River, and Padlei (1926) in the Barren Lands. As with the Indian tribes, fur economy changed the traditional Inuit ways. At Repulse Bay area, for instance, Inuit camps became sites of winter trapping, while sealing at the ice edge began to replace breathing-hole sealing as the main winter activity (Dumas 1968: 159).

By the late 1940s, the paternalistic approach of the Hudson's Bay Company was replaced by the Canadian wage/welfare system. This hybrid

Figure 3.3
Fur Trade Posts, *circa* **1760**

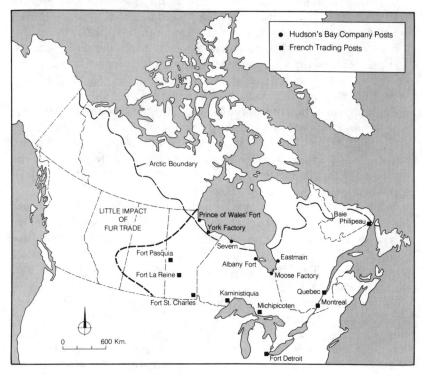

SOURCE: *The Canadian Encyclopedia* 1988: 858.

of European mercantilism and Native reciprocal-exchange traditions had worked well in the past but with the intrusion of trading rivals, missionaries, and governmental officials into the North, the old way quickly faded (Ray 1990: 221). Not surprisingly, the economic dependency so effectively mastered by the Hudson's Bay Company under George Simpson was 'inherited' by the Canadian Government.

Exploration and the Fur Trade

The fur trade had always been tied to exploration, as it searched for new trapping areas. Fur traders were eager to explore new lands and to make contact with distant Indians who might supply them with furs. Then, too, there was the need to reach the Indian hunters before a rival fur trader.

Fur traders occasionally undertook exploration for other reasons. In the case of Samuel Hearne's remarkable overland journey from Fort Prince of Wales at the mouth of the Churchill River to the Arctic Ocean in 1771–72, it was motivated by the desire to substantiate reports of a

Figure 3.4
Fur Trade Posts and Main Routes, *circa* **1850**

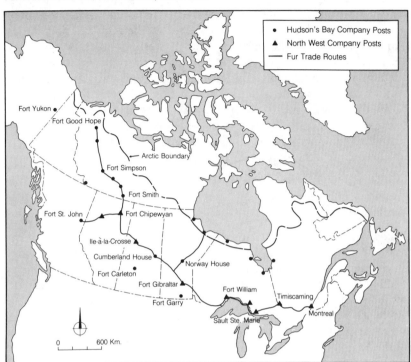

SOURCE: *The Canadian Encyclopedia* 1988: 859.

'rich' copper deposit along the arctic coast. In 1768, Northern Indians
brought several pieces of copper to the Hudson's Bay trading post of Fort
Churchill (Morton 1973: 291). The Governor of the post, Moses Norton,
sent Samuel Hearne in search of the deposit. In his first two attempts,
Hearne failed. Crossing the barrens was no easy feat, even with Indian
guides and company servants. Only with the help of Matonabbee and his
Northern Indians (including their women) was Hearne able to reach the
mouth of the Coppermine River in 1771 and examine the copper deposit
(Hearne 1959 and Rich 1958: 298). The success of this amazing journey
was due to Chipewyan knowledge of the land and animals. Hearne not
only reached the shores of the Arctic Ocean but he also travelled west to
Slave River. Near Great Slave Lake, Matonabbee traded European goods
for furs trapped by Dogribs which he would later exchange for more trade
goods at Fort Churchill (Morton 1973: 298).

Hearne made a number of important observations about the Chipewyans. For the first time, Europeans had some inkling of the enormous
extent of travel by Indians living in the Subarctic. Matonabbee and his

Chipewyans ranged a vast hunting territory, extending from Hudson Bay to the Arctic Ocean to Great Slave Lake. These forested and tundra lands, interconnected by rivers and lakes, were not the exclusive hunting grounds of the Chipewyans, nor were they marked by fixed boundaries. For example, during the summer, some of these Indians moved into the Barrens, which also served as the hunting territory of the Inuit. But in these sparsely populated lands, the chance of contact was slim and direct contact could easily be avoided. Yet, at this particular time in history, these Chipewyans were the dominant group on the Barrens, partly because of their leader and partly because they were armed with muskets and other European weapons. As middlemen in the fur trade, they were able to terrorize the Inuit and exploit neighbouring Indian tribes such as the Dogribs and still profit from trade at Fort Churchill.

Whaling in the Arctic

While most inland exploration was conducted by fur traders with the assistance of their Indian guides, English, Scottish, and American whaling ships penetrated the northern waters. In the eastern arctic, Baffin Bay and Hudson Bay were popular whale-hunting areas; in the western arctic, whaling ships followed the Alaskan coast to the mouth of the Mackenzie River. These whalers and the British naval expeditions of the early nineteenth century helped find the elusive Northwest Passage. The tragic loss of the Franklin expedition (Vignette 3.5) indicates the perilous nature of these voyages.

Vignette 3.5   The Search For Franklin

In 1845, a British naval expedition headed by Sir John Franklin set out to complete the Northwest Passage. Franklin needed to find a sea link from Barrow Strait to the arctic coast near the mouth of the Mackenzie River and from there to the Pacific Ocean. By 1848, Franklin's expedition was assumed lost and efforts to find him, his crew, and ships sparked the greatest search in Arctic history. In 1853, a Hudson's Bay Company search party headed by John Rae first learned of the fate of the Franklin expedition from local Inuit. Six years later, an expedition headed by McClintock and Hobson, and sponsored by Lady Franklin, was successful in finding the remains of part of the Franklin party near King William Island, and the only written account of the expedition.
SOURCE: Cooke and Holland 1978: 212–16.

Whaling began in Davis Strait in the seventeenth century. Later, whaling ships ventured further north and even into Lancaster Sound. Dutch, German, English, and Scottish whaling ships plied these arctic waters in summer voyages. Between 1820 and 1840, whaling reached its greatest intensity with over a hundred ships involved. By 1860, American ships appeared in Hudson

Bay, and, several decades later, in the Beaufort Sea. Wintering-over by whaling ships was most common in the more inaccessible areas such as Hudson Bay and the Beaufort Sea. By the time a sailing ship reached Hudson Bay or the Beaufort Sea, sea ice had begun to form, making whaling impossible. These ships, spending the winter months frozen in a sheltered harbour, would get an early start in the spring, leaving time to sail south before the beginning of the second winter.

The impact of the whalers on the Inuit was greatest in places where wintering-over was common. While the population of whalers was not great (perhaps 500 men annually wintered in the Beaufort Sea and 200 in Hudson Bay), this contact drew the Inuit into a new economic system. Engaged in the whaling industry as pilots, crewmen, seamstresses, and hunters, Inuit hunters, now armed with rifles, greatly increased their take of caribou in order to supply the whalers with game. In exchange for these services, the Inuit obtained trade goods. Contact with the whalers had its downside, counterbalancing economic gains. Excessive drinking some- times led to violence, but by far the most negative impact of contact with Europeans and Americans was exposure to new diseases. Epidemics of smallpox and other contagious diseases swept through contact Inuit communities, reducing their populations. Even so, the sudden disappear- ance of the whalers upon whom many Inuit had come to depend was a blow. The whaling industry saw its products lose favour with southern consumers as new products came into the market place. First, it was petroleum which, in the late nineteenth century, replaced whale oil as a fuel for lighting. Then, a few decades later, the market for baleen col- lapsed when steel products were substituted. By World War I, commercial whaling in the Arctic had ceased.

The Beginnings of the Canadian North

In 1867, Canada became a nation and, in 1870, Rupert's Land and other holdings of the Hudson's Bay Company were transferred to Ottawa. At one stroke of the pen, the fur monopoly of the Bay was broken and its southern, more temperate lands were opened up to settlement. The company did receive generous land allotments, some of which it sold to settlers. A decade later, the British government transferred to Canada the rest of its arctic possessions, the Arctic Archipelago (even though all the islands had not yet been discovered). By 1880, the geographic extent of Canada had been realized, though Newfoundland was to join Confeder- ation much later.

North of the agricultural fringe, little resource development took place, leaving the Native inhabitants free to pursue their traditional activities. Still, Canada began to wonder what riches it had inherited. In 1888, a Senate Committee on the Resources of the Great Mackenzie Basin investigated this question. Later, the Senate published a 'highly enthusiastic report on the potential for agriculture, fisheries, forestry, mining, and petroleum, setting the precedent for the optimistic and promotional tone that has continued to this day to pervade government pronouncements on northern resources' (Rowley 1978: 79). While the Senate report sparked considerable interest, the national priority was to settle the Prairies. Little commercial activity took place in the Subarctic and northern life continued to revolve around the fur trade and the Hudson's Bay Company.

But in the late nineteenth century, economic changes were to make the North an economic hinterland of Canada rather than a fur-producing area of the Hudson's Bay Company. First, the forest resources of the southern edge of the boreal forest in Quebec and Ontario attracted the attention of Canadian and American entrepreneurs. In a classical core/periphery relationship, the eastern subarctic forest industry received capital from the giant newspaper firms in major American cities; in turn, it supplied these firms with pulp which was processed into paper at plants near these metropolitan centres. In later years, paper was also produced at mills in the Subarctic. Second, in 1897 the Klondike gold rush in the southern Yukon began to transform that area into a mining frontier.

Beyond these two developments, the remainder of the northern landscape remained a wilderness committed to the fur trade and to hunting. Indian relations with fur traders continued but with a growing dependency. Since most of the North remained a 'wilderness', why is it that relations between Indians and fur traders changed? One factor is related to the collapse of the pemmican trade with the demise of the buffalo herds. A food substitute had to be found. At first it was the wildlife around the trading posts, a resource soon over-exploited and reducing the food supply for both traders and local Indians. A second factor was the growing pressure on fur and wildlife resources from whites. After World War I, many went North to seek their fortune as trappers. Anderson (1924: 332) described its effects: 'An influx of white trappers into regions normally visited only occasionally by wandering Indians or Eskimos means an immediate decimation of the game and fur resources. . . .' Another factor may have been a growing tendency for Natives to remain near fur posts. Indeed, after the arrival of the white trappers in the North, the competition for fur and game soon depleted animal populations, perhaps adding to Native tendency to spend less time on the land and not to journey too far from the trading post where food might be obtained. This form of 'dependency' on food from the traders and later the federal government was firmly rooted in the fur trade as early as the late nineteenth century (Ray 1974: 16).

New Challenges–Resource Development

In the decades immediately following Confederation, little attention was paid to 'developing' the North; it was rather a matter of holding on to this vast territory with a minimum of effort and cost. As Canada's first Prime Minister, John A. Macdonald was concerned that American homesteaders might migrate northward into the 'unoccupied' Canadian Prairies, thereby leading to their annexation to the United States. For this reason, his primary concern was to exert political control over the Prairies by building a railway from Ontario to the Pacific Ocean and then to settle these newly acquired fertile lands. The railway was completed in 1885, and the settling of the Prairies continued into the early part of the twentieth century. During all this time, most of Rupert's Land and all of the North-Western Territory and the Arctic Archipelago were left to the fur economy.

The first challenge to the fur economy occurred in Yukon. Traces of gold had been discovered in Yukon as early as 1866, but the famous Klondike find of 1896 triggered a gold rush. Two Chinook brothers, Skookum Jim and Tagish Charlie, and George Carmacks, a prospector married to their sister Kate, made the world's most famous gold strike at Bonanza Creek. As news spread to the outside world, thousands rushed to Yukon. According to Crowe (1974: 121), the coming of so many white men shattered the world of the Tagish, Tutchone, and other Indian tribes in Yukon. These Indians lost control of their traditionally occupied lands and became involved in the gold economy. Their participation varied from prospecting to packing supplies from the coast over the Chilkoot Pass (Crowe 1974: 122). Some found a place as wage-earners as deckhands on the river boats, or even as carpenters in Dawson. Hunters sold game to the miners.

The dark side of the gold rush saw the outsiders occupying Indian lands, killing wildlife for food, and exposing local people to 'new' diseases. Perhaps even more significant to the Indians was the sudden imposition of another economic life style, forcing major adjustments in their way of life. Some integrated as best they could while others attempted to continue their old ways.

In 1896, with the sudden arrival of so many newcomers in Yukon, the issue of Canadian sovereignty arose. Prior to the gold rush, Yukon had been inhabited by Indians and a handful of white traders and prospectors, and Ottawa had no need to send its officials there. Now Ottawa quickly dispatched detachments of the North-West Mounted Police to Yukon, their main purpose to enforce Canadian laws and regulations. The Mounties issued whaling licenses, collected taxes, and kept law and order. In this way, Canadian sovereignty was demonstrated. This was important because many of the miners and whalers were American citizens and fears of annexation were widespread in Ottawa.

The police also began to enforce Canadian laws on the Native peoples. For the most part, Indians and Inuit could avoid such interference with their ways by keeping away from settlements. However, concerns about wildlife mounted in southern circles and in 1916–17 Parliament passed two acts that greatly affected Native hunting activities. These were the Northwest Game Act and the Migratory Birds Convention Act. The police were obliged to apply these game laws to Native hunters, thereby adding another irritant to Native relations with federal officials.

Vignette 3.6 Changing Life in Yukon

Restless prospectors were exploring the world for more gold, and by about 1870, a few had entered Yukon by different routes. One man, George Holt, was able to cross the Chilkoot Pass, so jealously guarded by the Chilkoot Indians. An Alaskan Indian gave him two gold nuggets, and the stories he told after returning south brought more gold-seekers to the Yukon. Twenty prospectors, protected by a U.S. gunboat, met the Chilkoot chief Hole-in-the-Face, and by firing a few blank rounds from a machine gun forced the Chilkoots to open their mountain pass to all comers. From then on a steady trickle of prospectors toiled up the thousand-foot-high pass and into the Yukon River basin. The Chilkoot, Chilkat, and related Indians, having lost their control of the inland trade, now began to make money as packers, carrying loads over the pass as they had always done. By 1898, when the gold-seekers were desperate to cross the pass, the Indian packers could make a $100 a day and more, carrying loads of up to 200 pounds across the mountains.

SOURCE: Crowe 1974: 122.

Native People

In Canada, the term Native people refers to those Canadians whose ancestry includes at least one Indian or Inuit person who was living in North America at the time of European contact. Native people are very diverse and they form hundreds of linguistic groups, although most now speak English. Three of the larger groups are those speaking Cree, Inukti-tuk, and Slavey. Sometimes these linguistic groups comprise a regional population such as the Inuvialuit. In the case of the Métis, however, English or French may have been spoken rather than an Indian language.

Because of their special relationship with the federal government, Native people have been classified in four legal categories. These are Status Indians, Non-Status Indians, Métis, and Inuit. The distinction between Canadians of Indian ancestry (Status Indians, Non-Status Indians, and Métis) is not based on biological criteria but can be explained by their relations with Europeans and later with Canadians of non-Native ancestry. Out of these relations came legal definitions of these four groups which can be traced back to Confederation. At the time, the British North America Act assigned the federal government responsibility for 'Indians, and Land reserved for Indians'. In 1876, the Indian Act was passed and Indians were placed in a different legal position from other Canadians. The Inuit have never been subject to the Indian Act. In 1939, however, a court decision ruled that they were a federal responsibility. In 1982, the Métis gain official recognition as one of the three 'aboriginal peoples of Canada'. Then there are non-status Indians who, for one reason or another, have lost or surrendered their status.

Canadian Indians who have 'status' fall under the Indian Act. Those status Indians whose ancestors signed a treaty with representatives of the federal government have treaty rights. These treaties called on the Indians to surrender their aboriginal rights to vast tracts of land in exchange for reserves, the right to hunt and fish, ammunition and fishing twine, annual payments to each band member, and uniforms and medals for the chiefs. Under most treaties, the annual payment to a band member is $5.00. Those status Indians and Inuit who have not signed treaties can still negotiate a land claim with the federal government. Some may be comprehensive claims like that of the Inuvialuit of the western Arctic. In 1984, the Inuvialuit and the federal government agreed to a settlement. The Inuvialuit agreed to surrender to Canada all aboriginal claims in the Northwest and Yukon territories in exchange for hunting, fishing, and trapping rights; title to lands; and a cash settlement of $45 million in 1977 dollars.

The Métis people came into being during the early fur trade era. At that time, many fur traders took an Indian wife. Children of these unions did not all become Métis. Some remained with their mother's tribe while others melted into white society. Many, however, became identified as Métis. Over time, they have developed a separate culture and history.

Aboriginal Rights and Land Claims

Aboriginal Rights is the legal basis for Native land claims in Canada. Aboriginal Rights have yet to be legally defined but the term refers to 'property' rights which Natives retain as a result of their traditional and present use of the territory for hunting and trapping. A treaty between a tribe and the federal government is an agreement by which the tribe exchanges their interest in their ancestral territory in exchange for a relatively small amount of land called a reserve. The agreement also requires the government to fulfil a number of commitments described in the treaty. The basis of aboriginal rights is founded in British justice and was first expressed as a legal document in the Royal Proclamation of 1763. In this official declaration, ownership of land occupied by Indians is recognized. In the early 1760s the British and French were at war. Each European nation allied itself with Indians living on lands that it controlled or claimed. The British allied themselves with the Iroquois, and in turn the British government declared that the lands west of the Appalachians were to remain Indian lands. Settlers from the thirteen British Colonies were not to cross the Proclamation Line of 1763, a 'line' following the

Appalachian Divide from Maine to Georgia and then to the St Marys River in Florida. Efforts to prevent settlers from moving west proved futile, however, and by 1776 as many as 100,000 people may have been living west of the imaginary line (Hilliard 1987: 149). After the American Revolution, the thirteen states claimed the western territories (lands extending from the Appalachians to the Mississippi River) and thousands of American settlers poured over the divide. Within several decades, virtually all of the land east of the Mississippi had passed into white ownership and the original inhabitants were exterminated, assimilated, or living on reservations (Hilliard 1987: 163).

In British North America, the British treaties and political alliances with Indians were now directed against the Americans. This strategy ended after the war of 1812 when relations between the United States and Britain became less belligerent. After Confederation, Canada continued the tradition of making 'treaty' with the Natives. Treaties involved the surrender to the Crown of the property rights to lands traditionally used by Indians for hunting and trapping. In exchange for their surrender of aboriginal claims to the land, the Indians received reserves, lump-sum cash payments, an annual payment to each member of the band, and promises of hunting and fishing rights on unoccupied Crown lands.

While the exact terms varied somewhat, the basic elements of treaties were protection of the Natives by the Crown from the settlers and prospectors, and the establishment of their own land base. The land was granted to the tribe or band and ownership was a communal arrangement which could not be sold. In the case of the Métis, however, they received scrip which could be converted either to cash or land. For the most part, the scrip was then sold, and the Métis do not have a similar 'reserve' land base. The reserve land base represented a very small area compared to the Natives' former hunting grounds and while the federal government encouraged southern Indians to become agriculturalists, those in the Subarctic remained free to pursue a hunting and trapping life.

Some Indians in the territories have taken treaty and a few have selected reserves. In other cases, reserves have yet to be determined. For these territorial-based Indians and Inuit, and those living in British Columbia, another avenue is open, that of comprehensive land settlements. The Inuvialuit have made such a settlement and the James Bay Natives have a similar type of agreement. In both cases, a large land base and cash settlement have been agreed to in return for the surrendering of Aboriginal title to the land. The topics of comprehensive land settlement, treaties, and reserves are discussed more fully in a later chapter.

Treaty-making with the northern Indians is still not complete. Until the late nineteenth century, the federal government showed little interest in making a treaty with northern Indians who occupied non-agricultural lands. Then came the Yukon gold rush. The region was soon overrun by outsiders. Federal officials in Yukon reported this influx of southerners had negative social impacts on the Indians. In 1899 this commercial development and white settlement caused the federal government to sign Treaty No. 8 with the Indians and Métis south of Great Slave Lake. In 1921, shortly after oil was discovered at Norman Wells, Treaty No. 11 was signed with the Dene north of Great Slave Lake (Zaslow 1984: 8). In the 1970s, the Dene leaders stated that they did not accept the validity of Treaties 8 and 11, claiming that those Dene signing the agreement never agreed to surrender their land (Fumoleau 1973).

Resource development is the basis of the 'new' North and yet Canada's Native people find that they have been left out of this development process because they do not 'own' the land used by them. Without a land claims settlement, land used by Natives is actually Crown land, though Natives do have a right to hunt and trap on such land (but not on Crown land sold to individuals or companies). Settlement of land claims is not a simple matter because the federal and Native negotiators have been given widely different mandates. For instance, the federal government still treated mineral resources found on Crown land claimed by northern Natives as the property of the federal government. One exception to this rule is found in the Norman Wells oil production which is partly owned by the federal government. In this case, the federal government has declared that a share of federal oil profits would accrue to the Dene and Métis as part of a comprehensive land claims agreement.

The 'New' North

Near the middle of the twentieth century, the pace of resource development quickened, marking a major watershed in the course of northern history. Prior to that time, the history of the North evolved around hunting and the fur trade. The Inuit, for the most part, did not become involved in the fur trade until the twentieth century. During the Second World War, much of the North became a military defence zone. The old ways of living on the land changed too and the Dene, Inuit, and Métis were, for a variety of reasons, becoming settlement dwellers. As these events unfolded, a new cultural landscape emerged.

The 'new' North had become a vital strategic region. The North affected the Allies' war effort because it offered safe, inland supply routes to the European and Japanese theatres of war. Vast sums were spent to create a military infrastructure. While there was a lull in military activities after 1945, the North soon regained a geopolitical role as a buffer zone between the two superpowers–the United States and the Soviet Union. Now, it served an 'early warning radar' role for the United States which feared a surprise Soviet air attack. At roughly the same time, the demand for raw materials and energy rose and factories in the industrial heartland of North America turned to Canada's northern resources. Mega-projects became the order of the day. The iron ore development in northern

Quebec marks the beginning of such gigantic industrial projects. Watson (1963: 467–8) describes the transformation of a northern region from 'undeveloped to developed' and the link of such developments to American industrial needs:

> A railway was built from the little fishing village of Sept Iles–now a thriving port–up over the high, formidable, scarp-like edge of the faulted Shield, across extremely rugged terrain, deeply eaten into by rejuvenated sharply entrenched rivers, a distance of 360 miles to the Knob Lake iron field. Here the town of Schefferville soon grew up, a major outpost in the wilderness. The making of the port, the building of the railway, and the installation of near-by hydro-electric dams and works, were all major operations, but with American backing they were soon completed, and in 1954 the production of ore started, and over 2 million tons of ore were shipped out to the iron-hungry cities of the St. Lawrence and Great Lakes lowland.

Watson fails to mention the social impacts of this project on the local Native people–the Naskapi and Montagnais. Unlike the James Bay Project, no one raised the question of Aboriginal title to these yet unceded lands. The impact of these industrial, military, and governmental intrusions into the North has had an enormous impact on Aboriginal peoples. Post-World War II, Native society began a quick and rough transformation from a quasi-nomadic hunting life style to a sedentary one. In the past, Native people had lived on the land, coming into contact with few Euro-Canadians during the course of a year. By the 1960s, most Native people lived in settlements and had daily contact with other Canadians. As Abele (1989: 6) points out, 'each successive generation of Native people has grown up in very different circumstances.' Over three generations, the childhood experiences of Native people have differed sharply, illustrating the scope of social change and adjustment.

After 1945, a new phase in northern resource exploitation began. The production of minerals and forest products from the Canadian Subarctic grew rapidly, making the Subarctic an important source of raw material for American industry. As the economic value of the North became more apparent, Ottawa adopted new policies and programs designed to promote such development. In 1957, John Diefenbaker presented a northern development concept, the 'Northern Vision', calling for the opening of the northland by building transportation routes and communication lines, thereby linking northern resources to southern markets. In 1958, a series of programs was put into place, including 'Roads to Resources'. The 'Roads to Resources' program provided funds for new roads leading to potentially valuable natural resources. The federal share of these monies was determined by formula sharing costs with the provincial governments. The territorial counterpart was the Development Road Program with the federal government paying for all the construction costs.

During the next decade, northern development continued to hold Canadian's attention. In 1969, a more complex version of public involvement in northern development was proposed by Richard Rohmer. His Mid-Canada Development Corridor proposal called for the building of a northern railway across the mid-north, a transportation corridor meant to stimulate settlement and development along the lines of the building of the CPR across the Canadian West in the 1880s. In many ways, this grand scheme was an elaboration of Diefenbaker's Northern Vision. According to Rohmer, investment and planning would come from the federal government, ensuring Canadian control and ownership. Rohmer envisioned this massive federal undertaking as a means to strengthen both Canada's economy and its national purpose. He saw the Mid-Canada Corridor Concept as a counterweight to the American ownership of Canadian natural resources.

Canadians, particularly non-Native northerners, responded warmly to Rohmer's concept of a developed North. Federal officials, on the other hand, were cool to his idea of building a railway across the Subarctic, partly because of the size of the potential drain to the treasury but mostly because Ottawa considered it 'unsound'. In the west, a handful of provincial officials reacted suspiciously, fearing that this Middle Canada transportation system would serve the interests of Central Canada rather than those of the western provinces. The mixed response to this project indicated that while Canadians wished to see the North developed, there were differences of opinion how the goal was to be accomplished.

Environmental and native rights spokespersons were less enthusiastic and challenged development schemes. What were the anticipated impacts on the fragile polar environment? How would such industrial projects affect native land claims? These issues continue to stir social thought and political action in Canada and, in a broader sense, the process of resolving these critical issues is changing the conscience of Canadian society.

The Military in the North

The first military expenditures in the Canadian North took place during the Second World War, not to defend the northern territory but to develop supply lines to major theatres of war in the Pacific and Europe. The U.S. government funded most of the projects because it was the most effective way of supplying its troops in Alaska and England. The four main projects were (1) the construction of the Alaska Highway; (2) the building of a series of landing strips north of Edmonton to Fairbanks called the Northwest Staging Route; (3) the Canol Project which involved expanding oil production at Norman Wells and the laying of a pipeline from Norman Wells to Whitehorse and then to Fairbanks; and (4) the Crimson Staging Route which saw a number of landing strips built north of Winnipeg to Frobisher Bay and then on to Greenland and England.

One impact of military spending was to create a northern transportation infrastructure (Vignette 3.7). Landing strips were built in remote centres and the Alaska Highway provided the first road link from southern Canada north of the 60th Parallel. While this improved transportation system was designed to meet military needs, it also made northern resources more accessible to world markets. Gradually, more and more of the Subarctic became integrated into the market economy, supplying raw materials and energy to the populated areas of Canada and the United States. Like the impact of the permanent fur-trading posts of the Hudson's Bay Company in the late seventeenth century, towns based on primary industries and government services had an impact on the human geography of the North by extending a hierarchical network of trade centres across the North and thereby more closely integrating northern peoples and their activities into the Canadian economy and society.

A second effect of the military was the involvement of Native people in the wage economy and their subsequent movement into settlements. The Native workers found employment with the military an appropriate way to obtain western products, such as steel knives, rifles, and building materials, and southern foods, medical services, and entertainment. As the Native families became more and more attached to the construction camps, they became known as settlement Eskimos to differentiate them from those who continued to live on the land.

Military interest in the North changed during the post-World War II period. Until 1990, the United States saw the Canadian North as a buffer zone between it and its super power rival, the Union of Soviet Socialist Republics. As the military capabilities of the USSR increased, the United States improved its early-warning radar system. During the 1950s, three radar systems were built in Canada. These were the Pinetree Line (completed in 1954), the Mid-Canada Line (completed in 1957), and the Distant Early Warning Line (completed in 1957). The 22-station DEW Line was a massive construction effort that employed over 25,000 workers and cost well over $1 billion. In 1985, Canada and the United States agreed to build the North Warning System to replace the now ineffective DEW and Pinetree lines. With the end of the cold war, the strategic importance of northern Canada greatly diminished and the expensive military plans for northern Canada have been set aside. As a new world order is established, the need for military bases in the North declines. For instance, is the federal government still committed to its 1989 announcement that a military base will be established at the arctic mining community of Nanisivik when its ore body is exhausted in 1992?

The Modern North

The era of the modern North based on resource development began in the 1950s. The Diefenbaker government's political commitment to the

Vignette 3.7   Major Military Developments During World War II

During World War II, a number of large-scale construction projects were launched. These included the construction of a series of landing fields, the creation of a military base at Goose Bay, the expansion of oil production at Norman Wells and the building of a pipeline from Norman Wells to Whitehorse, and the construction of the Alaska Highway.

Project Crimson called for the construction of a series of landing fields, to allow Canadian and American military planes to fly over the Canadian North to U.S. military bases in the United Kingdom. Using a polar air route, short-range tactical aircraft could easily make the long journey by refuelling at land strips in the Canadian North, Greenland, and Iceland. Airports and fuel depots were built along an eastern route—Mingan, Fort Chimo, and Frobisher Bay (now Iqaluit)—as well as at Churchill and The Pas.

The Northwest Staging Route had a similar purpose. With the threat of a Japanese invasion of Alaska, a secure supply route to Alaska was considered essential—hence the construction of the Alaska Highway and the upgrading of the Northwest Staging Route, a series of airfields with paved runways which provided an air route from the United States to Alaska.

The military complex built at Goose Bay served as a major American air base during World War II. In 1941, the western end of Hamilton Inlet was leased to the Canadian government for 99 years by the Newfoundland governing commission. This small subarctic wilderness was selected by the Canadian military officials because, in sharp contrast to the airport at Gander, it was not troubled by fog and cloud and therefore had more than twice as many flying days. Both Goose Bay and Gander were strategic links in the North American chain of defence and in supplying over 900 warplanes to the United Kingdom.

The Canadian government was able to ensure that most facilities were handed over to Canada at the end of hostilities. For example, the United States agreed to construct, maintain, and operate the Alaska Highway until six months after the end of the war, at which time the Canadian section would become the responsibility of the Canadian government.

'Northern Vision' was coupled to two other factors. The first was the growing presence of state welfare in the North. Native people greatly benefited from these programs, though programs such as public housing altered their way of life. The second factor was the growing demand for forest products, minerals, and energy in the United States. These factors combined to change the North from a fur-trading economy to a resource one, and to change Native society from a nomadic, land-based one to an urban-based one. New resource-based towns and Native settlements sprang up across the North and transportation links were built between most of these towns and southern centres.

In the North, development of resources for distant markets made the construction of a road or railway necessary (Vignette 3.8). Roads constructed under the 'Roads to Resources' program of the federal government, tended to connect resource-base centres with a major city in a province, and, in this sense, extended the provincial economy further north. During the 1950s, the grandest expression of 'northern development' was the development of iron ore mines and hydroelectric power in northern Quebec and Labrador and the building of a railway from Schefferville to Sept-Iles. The purpose of this privately funded project was to supply much needed iron ore to United States steel plants.

Vignette 3.8 Roads and Resource Development

Roads are similarly indispensable to industrial progress, whether they are tractor-train trails or paved highways. Tractor trains are still widely used for a month or two each winter, notwithstanding their great expense. Several tractor routes became summertime roads and eventually gravelled or paved highways or railway lines. The winter trail of the twenties from Lac St Jean to Chibougamau became first a year-round road, then the Beattyville-St Felicien railway branch. Another running from the town of Peace River to Great Slave Lake was improved in the late forties into the all-season Mackenzie Highway as a federal/Alberta project and later was the route for the Great Slave Lake Railway. Similarly, the British Columbia government extended the provincial highway system to the Peace River district to provide the rest of the province with a link to that region and to the Alaska Highway and to give Peace River settlers their long-awaited direct outlet to the Pacific Coast. The John Hart Highway, built from Prince George through the Pine Pass to a junction with the Peace River district's road system after 1945, pioneered the route to be followed by the PGE.

Along with economic growth, a remarkable social change occurred among Native peoples. This change was marked by their settling into tiny communities. There had been earlier signs of the impending shift from a subsistence land-base economy to a village life-style and a mixed economy. One of the first was the withdrawal of the Hudson's Bay Company support services for Indians and Inuit. In face of stiff competition and the growing presence of the Canadian government in the North, the company no longer felt an obligation to buffer Natives from economic swings in the fur market place. A second sign was the involvement of Dene, Inuit, and Métis in military construction work during the Second World War. Many of these Native employees lived at or near the construction site. This trickle of people exposed to settlement-like life grew to a massive flow in the 1950s.

Such rural-to-urban movements are tied to the process of industrial development. In many Third World countries, large shanty towns have grown up around the major cities. The reasons for Third-World migration are complicated and somewhat different from those of Indians, Inuit, and Métis, who, for the most part, chose to settle near their traditional trading posts. While this move marked the end of a nomadic life style and the family hunting unit, it does not mean the end of hunting and trapping nor the beginning of a life of destitution as found in the shanty towns of Third World countries. Still, there are few job opportunities available in northern settlements for people with little or no formal schooling.

During the 1950s, many northern Natives move to settlements. The move, while encouraged by the federal government, was not well thought out. Rather, it was in part a response to the deteriorating caribou hunt, upon which the Inuit and Dene along the west coast of Hudson Bay depended for food. Then too, the promise of employment and the availability of Family Allowance payments, health services, and housing lured some to settlements. These opportunities drew Native people into the economic and social system prevalent in the rest of Canada. In some cases, there was a deliberate attempt to use programs to 'ensure' such integration. Family Allowance payments, for instance, required that parents send their children to the community school. Since mothers and children tended to remain in settlements, the extended family hunting unit soon disappeared. Now Native men formed the hunting unit and spent much less time on the land than before.

The emergence of the Chipewyan community of Black Lake in northern Saskatchewan demonstrates the swiftness of change (Bone, Shannon, and Raby 1973: 22). Prior to 1950, these Chipewyans hunted caribou and trapped fur-bearing animals. Once or twice a year, they would visit the Hudson Bay trading post at Stony Rapids; the rest of the year, they spent on the land, moving from one hunting area to another. In their economy, hunting took precedence over trapping. During the early 1950s, a Roman Catholic church was built on the shores of Black Lake and a few Chipewyan families built log cabins near the church. By 1956, over half of the Chipewyan families had established houses at Black Lake and their children were attending the newly constructed school. Family hunting parties were becoming less common because mothers and school-aged children remained in Black Lake. The decision to become permanent residents of Black Lake was influenced by many factors–proximity to their Church and the local Hudson's Bay Store, and the availability of federal health services and programs such as Family Allowance.

In the 1950s, other far-reaching population changes began to take place. One of these changes was the number of new resource towns springing up in remote areas of the Subarctic. Thompson was one of these 'planned' instant communities. Located some 740 km north of Winnipeg, Thompson soon became a major producer of nickel. Its population increased rapidly and stood at nearly 15,000 in 1986.

By the mid 1950s, a new North with an urban population pattern was emerging. Using the 1956 census data, Gajda (1960) identified three northern population zones. One was along the southern edge of the Subarctic where a mixture of agriculture, logging, and mining took place, and where most non-Natives lived. The Subarctic proper contained far fewer people; it was characterized by vast areas of forest dotted with mining settlements, lumber camps, and fur-trading posts. The Arctic was inhabited by Inuit, except for a few military settlements. Throughout these three zones a land-based economy existed, supplying Native people with food and cash. In most of the Subarctic and all of the Arctic the Native population formed a majority. Along the edge of the Subarctic, however, Native people were vastly outnumbered by other Canadians.

Native People's Place in Development

The spread of the resource economy into the North has resulted in conflicts over land. But this conflict is over more than land–it is a struggle between conflicting goals, preferences, and values. One goal, northern development, is to integrate the North and its people into the Canadian industrial society. This society reflects a set of values, preferences, and values common to most Canadians. The process of integration goes beyond strictly economic matters and involves the establishment of Canadian institutions and governments in the North. The other goal is less well defined but involves Native people establishing their place in Canadian society and the northern economy. Most Native leaders see this goal best achieved by the establishment of a series of political homelands for Aboriginal peoples. This approach requires the settling of many outstanding land claims. In the meantime, the ever-expanding resource economy is acquiring more and more Crown land, some of which is claimed by Indian bands. In the past, resource companies have not treated the environment well, nor have they respected land claimed by Native Canadians. Industrially polluted lands and waters have affected all northerners but particularly those of native ancestry who depend more on game and other wild foods than do other Canadians. Oil, mining, and timber leases of crown land are often obtained by resource companies and these leases sometimes contain lands claimed by Native people. The federal and provincial governments may have been at fault for granting these leases but their position has been that resource development should not be impeded by a land claim. These same governments have shifted their position somewhat over the last decade and now they may restrict development on leased land which is claimed by Indian bands.

One example of the complexity of the development process is found in northern Alberta where the Lubicon band is located in a resource-rich area. This band has not yet concluded a treaty with the federal government. The Lubicon band claims 10,000 km^2 in northern Alberta where oil production is occurring and where the Daishowa pulp mill has a timber lease. These developments have also led to the extension of highways which facilitate further petroleum and logging activities, and which have had a detrimental affect on the wildlife. Without title to land, the Lubicon cannot prevent these changes to the natural environment nor do they receive any of the resource revenues. The band has claimed around $170 million in compensation for oil and gas taken in the past (*Star-Phoenix* 1989: D14). Progress toward a land claim settlement has been slow. In reaction, the Lubicon has tried to block further developments. For example, the band negotiated an agreement with Daishowa in March 1988 that no logging would take place on lands claimed by the band until a settlement is reached. Two years later, however, a logging firm that has a contract with Daishowa wanted to begin logging on lands claimed by the Lubicon (MacDonald 1990: A7). While these kinds of conflict slow down economic development, force local firms to lay off employees, and may result in legal action, the costs are due to the unfinished business of land claims; another cost, of course, is the diversion of Native energies from focusing on their development. Clearly, unsettled land claims remain a fundamental problem facing Canadian society.

Selected Readings

Bryan, Alan Lyle, 1986. 'The Prehistory of Canadian Indians', in *Native Peoples: The Canadian Experience*, edited by R. Bruce Morrison and C. Roderick Wilson. Toronto: McClelland and Stewart, Chapter 3.

Duffy, R. Quinn, 1988. *The Road to Nunavut: The Progress of the Eastern Arctic Inuit since the Second World War*. Kingston: McGill-Queen's University Press.

Fumoleau, Rene, 1976. *As Long As This Land Shall Last: A History of Treaty 8 and Treaty 11, 1870–1939*. Toronto: McClelland and Stewart.

Harris, R. Cole, 1987. *Historical Atlas of Canada. Vol. I (From the Beginning to 1800)*. Toronto: University of Toronto Press.

Hickey, Clifford, G., 1986. 'The Archaeology of Arctic Canada', in *Native Peoples: The Canadian Experience*, edited by R. Bruce Morrison and C. Roderick Wilson. Toronto: McClelland and Stewart, Chapter 4.

Miller, J.R., 1989. *Skyscrapers Hide the Heavens: A History of Indian-White Relations in Canada*. Toronto: University of Toronto Press.

Morrison, William R., 1983. *A Survey of the History and Claims of the Native Peoples of Northern Canada*. Ottawa: Department of Indian Affairs and Northern Development.

Ray, Arthur J., 1974. *Indians in the Fur Trade: Their Role as Trappers, Hunters, and Middlemen in the Lands Southwest of Hudson Bay, 1660–1870*. Toronto: University of Toronto Press.

———, 1984. 'Periodic Shortages, Native Welfare, and the Hudson's Bay Company, 1670–1930', in *The Subarctic Fur Trade: Native Social and Economic Adaptations*, edited by Shepard Krech III. Vancouver: University of British Columbia Press: 1–20.

————, 1990. *The Canadian Fur Trade in the Industrial Age*. Toronto: University of Toronto Press.

Rich, E.E., 1967. *History of the Hudson's Bay Company, 1670–1870*. Vol. 2. London: Hudson's Bay Record Society.

Zaslow, Morris, 1984. *The Northwest Territories, 1905–1980*. Canadian Historical Association Historical Booklet No. 38. Ottawa: Canadian Historical Society.

————, 1988. *The Northward Expansion of Canada, 1914–1967*. Toronto: McClelland and Stewart.

II
THE PROCESS OF
NORTHERN DEVELOPMENT

Chapter 4

POPULATION GEOGRAPHY

Beyond the Canadian Ecumene

Some 1.5 million Canadians–fewer than 6% of the population–inhabit our polar lands, an area comprising about three-quarters of the area of Canada. The population geography of Canada can therefore be divided into two parts, a relatively densely populated 'southern' zone and a sparsely populated 'northern' one. The southern zone is associated with the agricultural-industrial core of Canada while its northern counterpart consists of the resource hinterland found in the Arctic and Subarctic regions of Canada. Within the northern hinterland, the population is unevenly distributed with over 95% living in the Subarctic. Moreover, a much higher proportion of the northern population consists of Canadians with a Native background than is found in southern Canada.

These three factors–small numbers, low population density, and a different ethnic character–mark the major demographic differences between northern and southern Canada. In the decade to come, change in the size and character of the northern population may be brought about by the interplay of three demographic variables: fertility, mortality, and migration.

The primary reason for the small northern population lies in geography. The location of the North within Canada makes it a high-cost area, limiting economic projects initiated by the industrial resource sector, and

Table 4.1
Population Size by Northern Areas of Provinces and Territories, 1986

Province/Territory	Population in 000s		Per cent
Territories		75,742	5.2
Yukon	23,504		
Northwest Territories	52,238		
Western Canada		422,557	28.9
British Columbia	225,590		
Alberta	97,631		
Saskatchewan	25,340		
Manitoba	73,996		
Eastern & Atlantic Canada		962,081	65.9
Ontario	434,060		
Quebec	473,326		
Newfoundland	54,695		
Total		1,460,380	100.0

SOURCE: Statistics Canada. 1987. *Population: Census Divisions and Subdivisions.* Catalogue 92–101. Ottawa: Minister of Supply and Services.

the cold northern environment is not suited for agriculture. These two factors, poor accessibility to large markets and the absence of an agricultural base, greatly restrict settlement in the North. Their impact on population size is most apparent in the Arctic where fewer than 60,000 people live.

Virtually all northern Canadians live in urban places where jobs, services, and business opportunities are found. These urban centres vary in size from cities and towns like Thunder Bay, Val-d'Or, and Yellowknife to smaller communities like Churchill, Nain, and Old Crow. Most northern urban places but particularly Native settlements are small, having populations of less than 1000 persons. In the Arctic, over three-quarters of the settlements recorded in the 1986 census had fewer than 1000 residents.

Resource development is another reason why northern Canadians live in urban places. This economy also accounts for the massive influx of southern Canadians into the North. Almost all of these newcomers live in the Subarctic, particularly in the large, more accessible centres linked to the south by a modern transportation system. Government and service activities in the larger centres provide more job opportunities. Again, many of the people working in these activities were recruited in southern Canada. Aboriginal Canadians, on the other hand, tend to live in smaller, more remote communities, though a growing number of young, better educated Native Canadians reside in regional centres. The smaller settlements are located near traditional hunting and trapping areas.

Figure 4.1
Northern Census Divisions by Number and Name, 1986

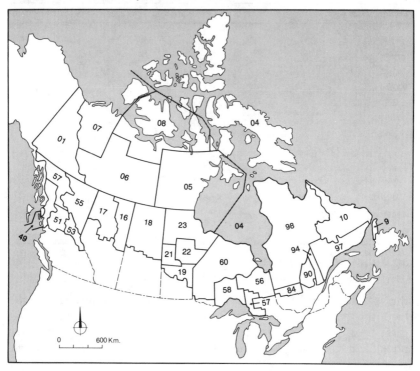

01	Yukon	21	Manitoba
04	Baffin Region, NWT	22	Manitoba
05	Keewatin Region, NWT	23	Manitoba
06	Fort Smith Region, NWT	56	Cochrane District, Ontario
07	Inuvik Region, NWT	57	Algoma District, Ontario
08	Kitikmeot Region, NWT	58	Thunder Bay District, Ontario
49	Kitimat-Stikine, BC	60	Kenora District, Ontario
51	Bulkley-Nechako, BC	84	Abitibi, Quebec
53	Fraser-Fort George, BC	90	Lac-Saint-Jean-Ouest, Quebec
55	Peace River-Liard, BC	94	Chicoutimi, Quebec
57	Stikine Region, BC	97	Saguenay, Quebec
16	Alberta	98	Territoire-du-Nouveau-Québec
17	Alberta	9	Newfoundland
18	Saskatchewan	10	Newfoundland
19	Manitoba		

The northern population is defined in terms of the census divisions situated in the Arctic and Subarctic. In 1986, there were 29 census divisions located in these two natural zones (Figure 4.1). The northern population of 1.5 million in 1986 is based on the number of people in these 29 census divisions.

Population Size

In 1986, the population in the 29 northern designated census divisions was nearly 1.5 million, forming nearly 6% of the national population. The population of each northern census division is found in Appendix 2 (page 244). The population size of each census division is a product of its physical landscape, economic factors, particularly limited access to local and global markets, and cultural attachment to the North, especially by Aboriginal peoples.

The exact population size of the North can only be estimated because the southern boundary of these census divisions does not exactly correspond with the southern edge of the boreal forest. For our purposes, however, the population of the North is that of these 29 census divisions.

Population Distribution

The population of the North is highly concentrated. Its uneven distribution can be examined from three perspectives. The first is the concentration of people in the Subarctic, rather than the Arctic. The second is the concentration of people in the eastern Subarctic–less than 30% of the northern population is found in the western provinces. The third is the concentration of people in regional centres. Commonly a regional centre is surrounded by 'empty' lands. This unique feature exits in all northern census divisions and is a reflection of both the northern economy and its physical base. In the large, sprawling Thunder Bay Census Division, for example, there were nearly 156,000 inhabitants. Approximately 98% of these people lived in or around the city of Thunder Bay. In sharp contrast, the northern part of this census division (north of the Canadian National Railway line) contained fewer than 2% of the total population. In the Thunder Bay District census divisions, the physical geography is dominated by the Canadian Shield. This rough, rocky landscape is ill-suited to agriculture (but suitable for logging, hunting, trapping, and fishing) and therefore most people live in Thunder Bay. This population pattern marks a fundamental characteristic of the northern land-use: most land is not occupied, but is used, particularly by resource companies and Native hunters and trappers.

The Arctic and Subarctic

The strikingly different population characteristics of the Arctic and the Subarctic are related to their varying economic capacities to support people. The Subarctic, where the resource base is more varied and accessible than that found in the Arctic, has well over 1.4 million inhabitants, while some 60,000 reside in the Arctic.

Within the Subarctic, most live along its southern edge in the so-called

Near North. Unlike the rest of the Subarctic, the Near North is integrated into the national transportation system. Here small cities, towns, and villages are based on resource exploitation for their economic well-being.

The Arctic, on the other hand, is the most sparsely settled region in Canada. This natural zone occupies most of the Northwest Territories and extends into northern Quebec, Labrador, Yukon, Manitoba, and Ontario. In 1986, it contained fewer than 60,000 inhabitants. Most resided in some forty arctic communities, ranging from Nain in Newfoundland to Tuktoyaktuk in the Northwest Territories. The bulk of the Arctic's population is found within four census divisions (Labrador, Baffin, Keewatin, and Kitikmeot). In 1986, these four units contained 47,452 people. Another 10,000 or so people are found in the arctic lands located in Nouveau Québec and the five other census divisions containing tundra lands: Yukon, Inuvik, Census Division 23 in Manitoba, Kenora, and Cochrane.

Another population difference between the Arctic and Subarctic is that most northerners living in the Subarctic reside in much larger urban centres than do their counterparts in the Arctic. Major metropolitan areas are Chicoutimi-Jonquière (158,486 in 1986), Thunder Bay (122,217), Fort McMurray (48,497), Timmins (46,657), Prince Albert (40,841), Rouyn (36,495), Corner Brook (33,730), Baie-Comeau (33,047), Alma (29,977), Sept-Iles (28,050), Val-d'Or (27,178), and Grande Prairie (26,471). To the north of these cities, there are a few large towns, such as Whitehorse (15,199), Thompson (14,429), and Yellowknife (11,753). Only very small settlements are found in the Arctic. Most contain fewer than 1000 inhabitants. In 1989, for example, only six of the 29 arctic communities in the Northwest Territories had populations of 1000 or more. At that time, the largest centre was Iqaluit with just over 3000 inhabitants.

A third difference is that the majority of people living in the Arctic are Inuit. Within the Subarctic, non-Natives form the majority in all census divisions except in Manitoba and Saskatchewan. In the subarctic of Saskatchewan three-quarters of the people considered themselves Canadians of Cree, Chipewyan, and Métis backgrounds. In the census divisions found in the Arctic, Inuit often form over 80% of the population.

Population Density

The population density of northern Canada is among the lowest in the world and is far below the Canadian average. Expressed as the number of people per unit of land area, Canada's 1986 population density figure was 2.8 persons per square kilometre and 0.14 for the North. The population density of Canada is roughly 20 times more than that of the North.

Population density figures are an indicator of the capacity of a region to support a given population at a particular standard of living. Land in

Table 4.2
Population Densities (people per km²) by Census Division, 1986

Yukon	0.04	*Manitoba*:	
		19	0.1
Northwest Territories:	0.02	21	0.5
Baffin Region	0.01	22	0.3
Fort Smith Region	0.04	23	0.04
Inuvik Region	0.02		
Keewatin Region	0.01	*Ontario*:	
Kitikmeot Region	0.01	Cochrane (56)	0.6
		Algoma (57)	2.8
British Columbia:		Thunder Bay (58)	1.4
Kitimat (49)	0.4	Kenora (60)	0.1
Bulkley-Nechako Regional District (51)	0.5		
Fraser-Fort George Regional District (53)	1.7	*Quebec*:	
Peace River-Liard Regional District (55)	0.3	Abitibi (84)	2.2
Stikine Region (57)	0.02	Lac-Saint-Jean-Ouest (90)	1.1
		Chicoutimi (94)	3.8
Alberta:		Saguenay (97)	0.5
16	0.4	Territoire-de-Nouveau-Québec (98)	0.05
17	0.3		
		Newfoundland:	
Saskatchewan:		Division 9	1.8
18	0.1	Division 10	0.1

SOURCE: Statistics Canada 1987/88. *Profiles: Census Divisions and Subdivisions.* Catalogue 94–101 to 124. Ottawa: Minister of Supply and Services.

northern and southern Canada varies widely in its capacity to support people, i.e., there are sharp differences between the two regions in soil fertility, locational advantages for industry, and access to large markets. Therefore, the population density figure for northern Canada does not mean that the Canadian North is underpopulated; rather, it indicates the capacity of the region to support a given population (Vignette 4.1). In general, the Subarctic can support more people than can the Arctic.

Within the North, population densities vary widely (Table 4.2). The lowest ones are found in census divisions located in the Arctic. These census divisions are Baffin Region, Keewatin Region, and Kitikmeot Region. They have densities below 0.01 person per square kilometre. Those census divisions with the highest density figures extend beyond the Subarctic into a mixed forested region where a major city is located. The census division of Chicoutimi, for example, has a population density of 3.8 persons per square kilometre. In comparison with the population density of 0.006 square kilometre for Kitikmeot District, the adjusted figure for Chicoutimi is still 200 times greater. This difference between the two census divisions is indicative of the variation in population density between the Subarctic and Arctic.

Vignette 4.1 Physiological Density

Variations in population density do not necessarily imply that there are too many people in one country or region compared to another one. The capacity of countries and regions to support a given population depends on many factors, including the local resource base. The Food and Agriculture Organization of the United Nations recognized this problem while comparing population densities of various countries. It attempted to devise a more meaningful measure by expressing population density by the amount of arable land per person. This measure is call a physiological density. For northern Canada, a physiological measure would have to be adjusted to recognize its non-agricultural land-use. Such a measure might well indicate that, given its natural land base and resource economy, the North is over-populated!

SOURCE: Newman and Matzke 1984: 34–5.

Population Before Confederation

Around 1860, the population of the North was likely about 60,000; we can only guess at the size of the northern population at earlier times. At the time of contact with Europeans, it might have been just over 100,000. What we do know is that this population consisted of Indians and Inuit, and a growing number of half-breeds who later became known as Métis.

The only comprehensive population estimates were produced by the Hudson's Bay Company for the British Colonial Office around 1850. It is widely accepted, however, that the Native population declined after initial contact with Europeans as they lost their hunting grounds and were subjected to diseases introduced to the New World by Europeans. When disease struck, the numbers of Natives in affected areas could drop by as much as half. Afterwards, the population might again increase. For these reasons the Native population in the North was subject to considerable fluctuations, though the general trend was downwards.

Population Changes after Confederation

Over the last 125 years, the population of the North has grown from around 60,000 to 1.5 million. This remarkable population increase was first due to the public policy of settling the newly acquired lands (Rupert's Land and the North-Western Territory) and later due to resource development. Initially, Canadians began to occupy the southern fringe of the Subarctic in Ontario and Quebec. This occupation was triggered by the building of the Canadian Pacific Railway which provided access to the timber and mineral resources of the Canadian Shield in Ontario and Quebec, and unplanned resource towns based on mining and logging began to appear. In 1871, the Census of Canada recorded a population of 3.7 million. Nearly all of these people lived in the Maritimes, the St

Vignette 4.2   The Balancing Demographic Equation Applied to the Northwest Territories

Measuring population change is a simple process involving three demographic factors, namely births, deaths, and migration, and the use of the 'balancing equation'. The formula for the balancing equation is $P_2 = P_1 + (B\text{-}D) + (I\text{-}E)$

where P_2 is the population to be estimated for the current year
  P_1 is the known population for a previous year
  B is the number of births taking place since P_1
  D is the number of deaths taking place since P_1
  I is the in-migration taking place since P_1
  E is the out-migration taking place since P_1

Applying data from the Bureau of Statistics of the Government of the Northwest Territories for the period extending from June 1987 to June 1988, a population estimate for June 1988 is made.

$$P_1 (1987) + (B \quad - D) \quad + (I \quad - E) \quad = P_2 (1988)$$
$$51,744 \quad + (1,505 - 209) + (3,548 - 4,485) = 52,103$$

SOURCE: GNWT Bureau of Statistics, *Statistics Quarterly*, September 1988 and June 1989: 5–7.

Lawrence Valley, and near Lake Ontario. This first Canadian census declared that 60,000 persons inhabited 'Labrador, Rupert's Land and the North-West'.

Toward the end of the nineteenth century, the population of the North showed signs of increasing through the in-migration of southerners. In 1896, for instance, the discovery of gold in the Klondike resulted in a short-lived increase in population of some 30,000 persons. During this same time, the rate of natural increase among Native people is believed to have been very low and may have actually declined. The result was a change in the ethnic composition of the northern population. Many Native people found themselves a minority in their traditional homelands. By 1941, the Native population likely comprised only one-quarter of the total population. Forty years earlier, Native peoples had made up close to two-thirds of the total population of the North. Numerous outbreaks of contagious disease were recorded and they took a heavy toll of Indian, Métis, and Inuit lives. When in 1928, for example, an influenza outbreak reached the Mackenzie Valley, 600 Indians or one-sixth of the Indian population of the Mackenzie District succumbed (Berger 1977: 145).

During the first half of the twentieth century, the northern population continued to expand because of the in-migration of southern Canadians. At the beginning of the twentieth century, the North had some 200,000 residents; fifty years later, this number had increased to 600,000. The driving force behind this large population gain was the utilization of northern resources, luring more and more southern Canadians into the

Table 4.3
Estimated Population Growth of Northern Canada, 1871 to 1986 (000s)

Year	Size	Year	Size
1871	60	1941	350
1881	60	1951	600
1891	80	1961	900
1901	100	1971	1,200
1911	130	1981	1,500
1921	150	1986	1,500
1931	250		

*These population totals for the North are based on the decennial census figures. Except for 1981 and 1986, they are estimates. Prior to 1981, different census subdivisions were used.

NOTE: Historical population statistics are, at best, rough estimates. Prior to 1951, the census figures are best considered approximations. From 1951 to the present, the census data for northern Canada are much more reliable.

Subarctic. Along the southern edge of the northern forest, but particularly in the Clay Belt of Ontario and Quebec, pioneer farmers settled. Others moved to logging and mining towns. During the same time, rich mineral deposits were discovered in the Canadian Shield near railway construction and several mining towns, such as Sudbury and Cobalt, sprang up. Logging also became an important economic activity in the virgin forests of Quebec and Ontario. These resource-based enterprises supported a much larger population than ever before and, by 1951, it reached nearly 600,000 people. Population growth continued until 1981 when it reached 1.5 million. During the 1980s the population ceased to grow.

In the 1980s, sharp down-swings in world commodity prices caused lay-offs and mine closings, and a net out-migration. In the early 1980s, out-migration occurred in northern Quebec and Labrador where iron mines were closed or their operations reduced; in northern Ontario, where the size of the mining and forestry workforce diminished; and in northern Saskatchewan where a uranium mine was shut down. These closures resulted in regional population losses and a substantial outflow of people to southern Canada.

A slowdown in in-migration in the 1970s turned into an out-migration. The chief reason was a depressed resource economy. Only a high rate of natural increase among Native northerners prevented a more serious population drop in the 1980s. During the 1980s, for example, the average rate of natural increase of northern Native people was nearly triple that found in southern Canada. The high rate of Native natural increase has been remarkably stable and it now plays a key role in northern population equation.

Over the past 125 years, the northern population has increased from 60,000 to around 1.5 million. More recently, this population growth has stalled. Does this mean that the northern population has reached a plateau? Or is it a temporary respite? This topic is explored next.

Table 4.4
Northern Population Changes by Provinces and Territories, 1981–1986

Province/Territory	1981	1986	Per cent Change
Newfoundland	57,056	54,696	−4.1
Quebec	487,943	473,326	−3.0
Ontario	443,826	434,060	−2.2
Manitoba	73,668	73,996	0.1
Saskatchewan	25,304	25,340	0.1
Alberta	87,636	97,631	11.4
British Columbia	227,556	225,590	−0.9
Yukon	23,153	23,504	1.5
Northwest Territories	45,741	52,238	14.2
Northern Canada	1,471,883	1,460,380	−0.8

SOURCE: Statistics Canada 1987. *Population Census Divisions and Subdivisions*: Catalogue 92–101. Ottawa: Minister of Supply and Services.

Recent Population Changes

In 1986, the population of the North was nearly 1.5 million (Table 4.3). During the previous five years, the population of the northern census divisions had decreased slightly, from 1.47 million in 1981 to 1.46 million in 1986 (Table 4.4). The main factor in this slightly negative rate of growth was a net out-migration of people from the North.

During these five years, population change varied among the 29 northern census divisions. A summary of these changes reveals that the greatest declines took place in Newfoundland (Labrador), Quebec (Sagueney) and Ontario (Kenora District) while the Baffin Region of the Northwest Territories recorded the largest percentage increase (Table 4.4).

Why did the North suffer a population decline, and could this decline continue? Prior to the 1980s, attractive job opportunities in the resource and public sectors drew many southern Canadians to the North, especially the Subarctic. Most were attracted to jobs in new mining towns such as Schefferville, Faro, and Thompson and to the territorial capitals of Whitehorse and Yellowknife. By the late 1970s, fewer new jobs were being created in the public sector and jobs in the resource sector were declining. The major transfer of public service jobs from Ottawa to Whitehorse and Yellowknife had been completed by the 1970s, and the global demand for mineral and forest products declined in the early 1980s. Industry also had to face a demanding environmental and social assessment process. From industry's perspective, this review process increased the cost of doing business in the North; from the public's perspective, it provided a check on industry. Following the review of the Mackenzie Valley Pipeline Project proposal in the mid-1970s, large-scale industrial projects have had to undergo similar public scrutiny. These reviews often request modification of the project. The net result has been to slow down resource development in the North.

The combination of these factors first caused a halt in net in-migration and then led to a net out-migration. From 1981 to 1986, the out-migration resulted in a small decline of 11,000 in the northern population. This relatively small decline masks the impact on resource towns, particularly in the iron-ore mining region of northern Quebec and Labrador. Offsetting these local losses were substantial gains in the Northwest Territories and the northern parts of Alberta. In northern Saskatchewan, the population loss caused by the closing of a major mining community, Uranium City, was offset by a high rate of natural increase among Native people, leaving Saskatchewan with an overall increase of 0.1% for the five-year period. In northern Alberta, the continued growth of the oil sands city of Fort McMurray added to its population increase which reached 11.4%.

The North's population may again expand and such increases could begin in the 1990s. Such population gains require that out-migration slows and Native northerners continue to have high annual rates of natural increase. A return to past high levels of population growth seems unlikely. The resource industry, which was the main driving force behind past population growth, has become more capital-intensive, implying that fewer workers will be required in the 1990s than in the previous decade. Another factor is the growing tendency for mining firms to fly their workforce from major cities in southern Canada to the work site, rather than building resource towns in the North. Finally, the global rationalization of the resource industry does not augur well for an increase in the North's population because it pits low-cost labour in Third World countries against high-cost labour in northern Canada.

Natural Increase

In 1986, the natural rate of increase in northern Canada was 1.3% while the national average was just under 0.8%. This annual rate is based on a surplus (or deficit) of births over deaths, expressed as a percentage of the mid-year base population. In the case of northern Canada, there were 26,525 births and 8,333 deaths, leaving a net increase of 18,192. The rate was determined by dividing the 1986 population of the 29 designated northern census divisions (1,460,380 persons) into the net increase (18,192), giving a rate of 1.3%. The largest increase occurred in the Keewatin District of the Northwest Territories (3.5%) while the smallest one took place in Thunder Bay census division of Ontario (0.7%). The natural rates for each census division found in the North are shown in Table 4.5.

Changes in the northern vital rates have followed the first three phases of the Demographic Transition Theory (Vignette 4.3). The fourth phase calls for fertility (now around 18.2 births per 1000 persons in the North)

Table 4.5
Natural Rate of Increase by Northern Census Divisions, 1981–86

Northern Census Divisions	Births Number	Births Rate	Deaths Number	Deaths Rate	Increase Number	Increase %
Yukon	483	20.6	113	4.8	370	15.7
Northwest Territories:						
Baffin Region	345	34.6	65	6.5	280	28.1
Fort Smith Region	565	22.5	87	3.5	478	19.0
Inuvik Region	252	30.0	32	3.8	220	26.2
Keewatin Region	197	39.5	25	5.0	172	34.5
Kitikmeot Region	147	39.2	26	6.9	121	32.3
Place Unknown	1					
Total	1,507	28.9	235	4.5	1,272	24.4
British Columbia:						
Kitimat-Stikine Regional District (49)	816	20.7	167	4.2	649	16.4
Bulkley-Nechako Regional District (51)	734	19.6	173	4.6	561	15.0
Fraser - Fort George Regional District (53)	1,609	18.0	381	4.3	1,228	13.8
Peace River-Liard Regional District (55)	1,180	20.6	254	4.4	926	16.2
Stikine (57)	36	17.8	10	5.0	26	12.9
Alberta:						
16	1,018	20.9	129	2.6	889	18.2
17	1,272	26.0	244	5.0	1,028	21.0
Saskatchewan:						
18	942	37.2	147	5.8	795	31.4
Manitoba:						
19	320	35.0	83	9.1	237	26.0
21	463	19.2	154	6.4	309	12.4
22	898	29.4	124	4.1	774	25.3
23	279	27.2	55	5.4	224	21.8
Ontario:						
Cochrane (56)	1,500	16.0	690	7.4	810	8.6
Algoma (57)	2,043	15.5	897	6.8	1,146	8.7
Thunder Bay (58)	2,261	14.5	1,253	8.1	1,008	6.5
Kenora (60)	1,273	24.1	405	7.7	868	16.4
Quebec:						
Abitibi (84)	1,519	16.1	564	6.0	955	10.1
Lac-Saint-Jean-Ouest (90)	1,022	16.2	378	6.0	644	10.2
Chicoutimi (94)	2,272	13.0	1,019	5.8	1,253	7.2
Saguenay (97)	1,407	13.5	472	4.5	935	9.0
Territoire-de-Nouveau-Québec	874	23.5	154	4.1	720	19.4
Newfoundland:						
9	373	14.4	140	5.4	233	9.0
10	424	14.8	92	3.2	333	11.6

SOURCE: Statistics Canada, 1988. *Principal Vital Statistics by Local Areas 1986*. Catalogue 840–542 occasional. Ottawa: Minister of Supply and Services.

to fall to the same level as the northern mortality rate, currently 5.7 deaths per 1000 persons. Vital statistics records for the Northwest Territories are much more complete than for the northern provinces and they provide annual birth and death figures for 65 years. Over this long period, two major patterns are revealed. In the pre-1945 period, the birth and death rates were usually under 20/1000 but the death rate occasionally rose sharply and sometimes exceeded the birth rate. The effect of the Mackenzie District influenza epidemic in 1928, for example, accounts for a negative rate of natural increase in the Northwest Territories from 1927 to 1929.

After 1945 northerners, particularly Native northerners, had greater access to modern medical services. For Native mothers, it meant giving birth in a hospital or nursing station where the chances of both mother and baby surviving increased greatly. The birth rate rose sharply, reaching annual rates of 40 births or more per thousand people. During the 1960s, these extremely high annual birth rates declined but then stabilized at levels well above the national figure. Since 1973, the annual birth rates have remained above 27 births/1000 persons, indicating that the expected final phase of the Demographic Transition Theory (Vignette 4.3) has stalled. The socio-economic factors supposed to cause the final phase decline are apparently not functioning in the Northwest Territories. Does this mean that the population of the Northwest Territories will continue to exhibit a high fertility rate until economic conditions for the Natives improve?

The notion of a fertility threshold (below which birth rates do not fall) is given credence by stable Inuit birth rates over the last twenty years. Federal officials who expect a continuing decline in fertility among the Inuit are probably making the assumption that a change in attitude toward the family size is occurring as it does in the last phase of the Demographic Transition Theory (Hagey, Larocque, and McBride 1989: 6). But that theory is based on the premise that economic change has occurred and family incomes are rising as a result of economic growth. No such growth– or jobs–exist in the Northwest Territories for Native peoples. If a decline does occur, it will have to be for other socio-economic reasons. On the other hand, the continuing high birth rate may well be linked to low incomes, cultural factors (such as the extended family and the acceptance of children born out-of-wedlock), and public support programs (such as financial support for single mothers and family allowances).

While the birth rate in the Northwest Territories remains high, the death rate in the Northwest Territories dropped sharply as medical services became available, preventing serious epidemics and reducing the rate of infant mortality. By the early 1960s, it had reached levels approximating the national average.

The combination of high birth rates and low death rates in the Northwest Territories has led to a population explosion in the post-1945 period.

Figure 4.2
Population Change in Northern Canada, 1871 to 1991

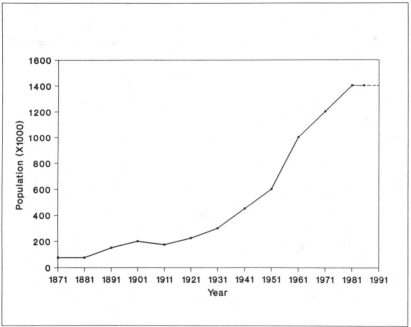

SOURCE: Censuses of Canada.

Since 1953, rates of natural increase have varied between 2.2% and 3.9%. Such figures are common to Third World countries but not industrialized ones. With few exceptions, these high rates of natural increase are associated with the Native population. For example, the Report on Health Conditions in the Northwest (1987: 22) reported that the Keewatin District of the Northwest Territories had both the highest proportion of Native people (88% in 1987) and the highest rate of annual natural increase (3.4% in 1987). In sharp contrast, those census divisions with less than 5% of their population consisting of Native peoples–Cochrane (4.5% Native), Abitibi (3.7% Native), Lac Saint-Jean Ouest (3.7% Native), Northern Peninsula of Newfoundland (1.4% Native) and Chicoutimi (1% Native)–with the exception of Lac Saint-Jean Ouest, all had natural rates of increase below 1% in 1986.

Migration

The migration of southern Canadians and foreigners to the North has been large enough and persistent enough to dominate the pattern of population, alter its demographic structure, and affect its ethnic composi-

Vignette 4.3 Demographic Transition Theory

The Demographic Transition Theory provides a basis for interpreting major changes in population growth at a global level. It describes demographic, social, and economic events which took place in Europe over several centuries. These events are linked to the process of industrialization which originated in western Europe. Its essential argument is that factors associated with a rise in the economic well-being of a society lead to a reduction of mortality and, after a short time lag, a drop in fertility. During these demographic changes, there is a large increase in population but the final phase culminates in little or no natural increase, i.e., zero population growth.

In northern Canada, there is no doubt that improved access to medical services accounts for the drop in the death rate in the post-World War II decades. Similar declines in mortality have taken place in most Third World countries. Yet the second vital decline, that of birth rates, has not occurred at the same rate in most Third World countries nor in northern Canada. In both cases population rates of increase remain very high. Since the Demographic Transition Theory calls for such a drop in birth rates to occur when incomes rise substantially, the existing low incomes for Native families may have the opposite effect. It stands to reason that, if the Demographic Transition Theory is correct, Native birth rates are unlikely to decline to levels approximating Native death rates until well after their income levels rise sharply.

SOURCES: Weinstein 1976, and Newman and Matzke 1984.

tion. These three impacts have occurred throughout northern Canada but vary by region. Most of the population of Yukon, for example, is composed of newcomers, that is, whites who have moved to Yukon over the last 100 years. As a result, Yukon has a much larger older population, i.e., most of its population is between the ages of 15 and 64 years of age, and non-Native residents predominate. Northern Saskatchewan, on the other hand, has been less affected by in-migration. For example, the proportion of Native people in northern Saskatchewan is around 75% compared to about 20% in Yukon.

Migration within a country usually takes place as people in one region react to the existence (or perception) of more favourable socio-economic conditions in other regions. In this 'push-pull' model, migration increases as socio-economic variations across a country widen. In the Canadian North, economic opportunities (high wages, bonuses, and living allowances) in the construction, public, and resource sectors have attracted many southerners. Another form of 'economic' migration occurs when companies and governments transfer employees from one region to another one. In this economic model, movement also occurs from rural areas to urban centres, on the assumption that urban areas offer more economic opportunities and social amenities than do rural areas.

Following World War II, Native people moved from the land to settlements, a migration beginning in the late 1940s and lasting for nearly two decades. Settlement living allowed greater access to trading posts, churches, and public services and programs but still permitted access to traditional hunting grounds. By 1970, nearly all Native peoples lived in a settlement. This migration transformed the population geography of the North and thrust Native people into an urban environment controlled by southern institutions and regulations. Federal government policy encouraging a relocation to settlements was based on the assumption that settlement life would facilitate the process of 'modernization', including sending children to school. Many of these settlements were former trading posts and very few have an economic base that can provide many employment opportunities. The lack of employment opportunities has not yet caused these northern Native Canadians to move to other regions of Canada because of their strong tie to 'place', including the desire for access to traditional land and family, and because few have the educational skills to find jobs in larger centres.

Southern Canadians tend to be 'economic' migrants. Many view northern migration as a two-way street. While non-Natives have moved to the North for economic reasons, they remain highly mobile and if an attractive job opened up elsewhere, they are likely to move (Vignette 4.4). This tendency is reinforced by the high degree of job uncertainty common to all resource industries whether they be mining, forestry, or oil exploration. During times of expansion, there is an in-migration of workers but, during an economic contraction, the same workers may seek employment outside the North. There is, therefore, an ebb-and-flow character to non-Native residents of the North, suggesting that 'transplanted' northerners have no intention of remaining but are simply saving their money to establish themselves in the south. The most extreme example was the Klondike gold rush. In 1898, some 40,000 came to Yukon, seeking gold and fortune (Stone 1989: 95). By 1898, Dawson City had a population of 18,000 but by 1920 it had dropped to 1200 (Crowe 1974: 126). In regional centres that have a more diversified economic base and many of the amenities of urban places, there is much more likelihood of southern migrants taking root and making the North their home. In such communities, government is usually the major employer, offering a wide variety of permanent jobs and the opportunity for an individual to advance within the system. This kind of economic stability plays a key role in keeping southerners in the North and eventually converting them to 'northerners'. That migrants change their attitudes to a new region over time is not a new concept to geographers. The existence of regional loyalties in Canada is readily acknowledged. Widespread agreement on regional issues is a visible sign of such loyalties. A territorial example would be the desire in both Yukon and the Northwest Territories for provincehood and a seat at First Minister Conferences.

Table 4.6
Net Migration between 1976 and 1981 by Northern Census Division

Northern Province/Territory	Net Internal Migration
Yukon	−550
Northwest Territories	−2,055
British Columbia	8,795
Alberta*	12,035
Saskatchewan	700
Manitoba	−9,325
Ontario	−5,700
Quebec	−16,515
Newfoundland	−4,685
Net Migration	−16,350

*Only Alberta census division 12; does not include the figure for Alberta census division 15 because it includes the heavily populated Peace River country.

SOURCE: Statistics Canada, 1983. *Population: Mobility Status*. Catalogue 92–907, Table 5. Ottawa: Minister of Supply and Services.

The strength of the out-migration from northern Canada is revealed by data from the 1981 census, comparing the residential locations of people in 1976 and 1981. During this five-year period, just over 16,000 more people left than entered the northern census divisions (Table 4.6). Two-thirds of the northern census divisions lost population. Gains were recorded in northern Alberta, British Columbia, and Saskatchewan — notably Alberta. Major losses occurred in Labrador, Saguenay, Nouveau Quebec, Kenora, Division 19 in Manitoba, and the Kitimat-Stikine District of British Columbia.

Vignette 4.4 The Best Reason to Come North . . . or Is It?

I've often asked myself why, four years ago, I jumped at the chance to work in the Arctic. In all honesty, I needed a job and the money was good. Heck, the money was excellent. And I think that's why most southerners come here. Our curiosity about the North is dwarfed by our thoughts of compound interest. But for the people who stay, it changes. Perhaps it's because we get used to food prices that are double and triple those in the South and money becomes less valuable. Perhaps our fascination with money is replaced by a fascination with the land and the people who were here long before we ever heard of the place. Perhaps one day, people will be willing to come to live in the North to seek that fascination. Perhaps one day, there will be no need to entice people north with high salaries, big benefits and two free trips a year. But let's not rush things.

SOURCE: Petrovich 1990: 41.

Demographic Structure

The demographic structure of a population–its age and sex composition–is measured in age cohorts, usually based on five-year intervals. Since the population processes of fertility, mortality, and migration shape a population over time, the demographic structure provides a profile of a particular population for a particular instant in time. The northern demographic structure is dominated by the youthfulness of its population. This youthful nature is due to the high northern birth rate and the out-migration of older non-Natives, retirees to warmer climes. In 1986, in almost all northern census divisions over 25% of their population was under the age of 15. The national average is 21.3%. Nearly half of the northern census divisions had 30% or more of their population under 15 years of age. The five census divisions with the highest percentage were northern Saskatchewan (38.7%), Baffin Region (38.6%), Manitoba Census Division 19 (38%), Kitikmeot Region (37.5%), and Manitoba Census Division 22 (36.5%). Only in northern Ontario and Quebec is an older population found. With the exception of Nouveau Quebec, no census division has 30% or more of its population under the age of 15.

Resource regions and towns tend to have two other characteristics: a large number of people of working age, and a predominance of males. The demographic structure found in Yukon exhibits these 'frontier' characteristics. In 1986, the birth rate of 20.6 births per 1000 persons was well above the national rate of 14.7 births per 1000 persons. Similarly, its sex ratio, a measure of sex composition of a population, was 110.4 indicating that there were 10.4 more males for every 100 females. In the early history of Yukon the population was much more male-dominated.

In the first half of the twentieth century, the male-dominated population reflected the mining economy and the frontier nature of Yukon. The 1986 age composition of Yukon's population revealed a high proportion of its population was under 15 years of age (25.1%) and between the ages of 25 to 44 (39.4%); while those over 64 form a relatively small percentage. For example, in 1986, 10.7% of Canada's population was over 64 years of age while Yukon had only 3.7%. A comparison of age groups for Yukon and Canada for 1986 is found in Figure 4.5. Similar demographic characteristics are found in other northern areas. In the Northwest Territories, for example, the population is extremely youthful, with 32.8% of the 1986 population under the age of 15, compared to 21.3% for Canada and 25.1% for Yukon.

Urban Population

Almost all northerners live in urban centres; in fact, the North has a higher percentage of urban dwellers than any other region of Canada.

Figure 4.3
Population Density by Northern Census Divisions, 1986

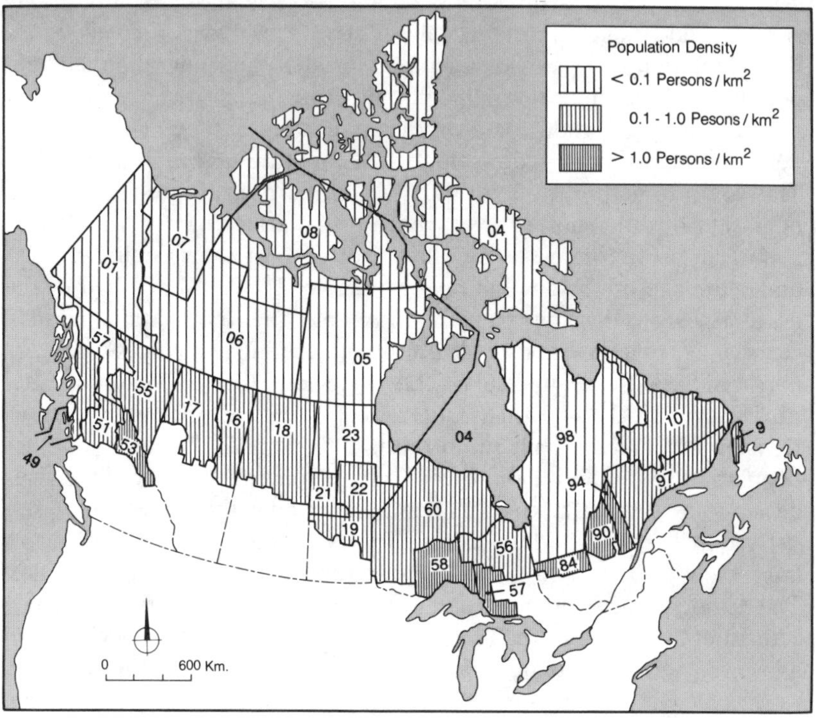

SOURCE: Census of Canada, 1986.

These urban centres form three types of communities: regional service towns, resource centres, and Native settlements.

Regional service centres provide public and private services, some of which are not found in smaller places. This fact results in a hierarchical arrangement of urban places. This spatial arrangement is described in central place theory, and involves the concept of threshold. For a firm to offer a set of goods at a particular place, it must sell enough to meet operating costs. The minimum level of demand that will allow a firm to stay in business is called the threshold. The same argument can be made for public services, i.e., a hospital and high school are located in regional centres while nursing stations and primary schools are found in smaller centres. On Baffin Island, Iqaluit with over 3,000 inhabitants in 1989 is the largest centre and has both a hospital and a high school. Lake Harbour with only 350 residents in 1989 is the smallest settlement on Baffin Island. Students from Lake Harbour who wish to graduate from high school would travel or move to Iqaluit, as would persons requiring medical care. The same urban hierarchical arrangement exists across the North. Inuvik

Figure 4.4
Changing Demographic Rates, Northwest Territories, 1924–1956 and 1957–1989

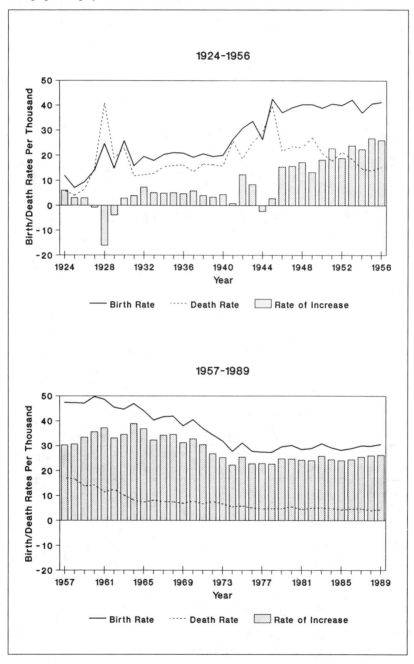

SOURCE: Statistics Canada, *Vital Rates*, cats. 804–204 and 206.

Figure 4.5
Broad Age Groups for Yukon and Canada, 1986

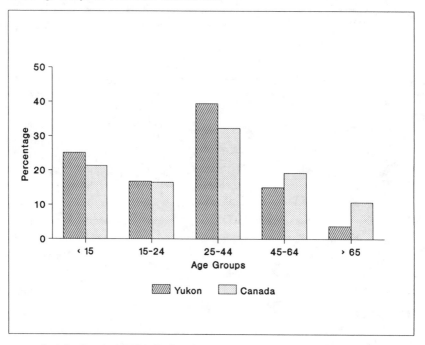

SOURCE: Statistics Canada, 1987/88. *Profiles: Census Divisions and Subdivisions.* Catalogues 94-101 to 124. Ottawa: Minister of Supply and Services.

Table 4.7
Changes in the Sex Ratio, Northwest Territories and Canada, 1911 to 1986

Year	Yukon	Northwest Territories	Canada	Difference Yukon	NWT
1911	325	106	113	212	−7
1921	211	107	106	104	1
1931	201	116	107	104	9
1941	179	125	105	72	20
1951	150	130	102	48	28
1961	127	126	102	25	24
1971	117	111	100	17	11
1981	111	110	97	14	13
1986	110	110	97	13	13
1987	109	110	97	12	13
1988	109	109	97	11	11

SOURCE: *Canada Year Book* 1930, 1956, 1975, and 1989.

with a population of 3500 is the regional centre for the Beaufort Sea-Mackenzie Delta area while Prince George with a population of nearly 70,000 is the regional centre for north-central British Columbia. Like Inuvik, Prince George provides education services to local and surrounding residents but it will soon house a university. By 1993, Prince George's new university will serve the population of north-central British Columbia, allowing it to become an education centre for the region.

Resource towns accounted for much past population growth. Now they are the principal source of out-migration. Single-industry centres have declined because of the depressed state of their principal industry. A recent example of a single-industry town losing both its mine and its people is provided by Pine Point in the Northwest Territories. In 1987, Cominco shut down operations in the lead-zinc mines at Pine Point and by 1990 Pine Point's population had dropped from nearly 2000 to under 100. Similar closures occurred at Schefferville and Uranium City.

Native settlements are growing and most of the population increase is due to natural increase. Since the highest rates of natural increase are found in the Arctic, most of these centres are growing very rapidly. Arctic communities that have become regional centres have also benefited from in-migration from smaller centres. Three such centres–Iqaluit, Rankin Inlet, and Cambridge Bay–increased their populations by over 20% from 1981 to 1986.

Northern settlements have a number of characteristics common to those in developing countries. One such characteristic is urban primacy, the concentration of the total population in one urban centre (Vignette 4.5). Yukon provides an extreme example of urban primacy. Its largest urban place, Whitehorse, with a population of just over 15,000, formed 65% of Yukon's 1986 population. The next largest town, Dawson City, had a population of just under 900.

A second characteristic is the large number of very small settlements. In the case of northern Canada, most settlements have fewer than 1000 inhabitants. In 1986, approximately 70% of 'urban' centres in the North had fewer than 1000 inhabitants. Many of these small settlements are located in remote places, on reserves, or at the sites of former trading posts.

A third characteristic is a rapid growth rate of Native settlements, particularly those in the Arctic. Even though most Arctic centres are small, many are increasing in size due to the high growth rate among the Inuit population. In 1961, only one (3%) community of the 29 urban places in the Northwest Territories' Arctic had a population over 1000 while by 1989 this figure had increased to seven (24%). Given the continued high rate of natural increase of the Inuit population, another five communities (Cape Dorset with 967 inhabitants in 1989, Coppermine with 945, Igloolik with 937, Pond Inlet with 909, and Tuktoyaktuk with 970) may reach a population of 1000 in the early 1990s. This growth rate

Table 4.8
Population Change in Arctic Communities, N.W.T., 1961 to 1989

Community	1961	1989	Change	% Change
Aklavik	599	762	163	27.2
Arctic Bay	49	554	505	1030.6
Arviat (Eskimo Point)	168	1,321	1,153	686.3
Baker Lake	386	1,081	695	180.1
Sanikiluaq	169	488	319	188.8
Broughton Island	70	473	403	575.7
Cambridge Bay	531	1,065	534	100.6
Cape Dorset	161	967	806	500.6
Chesterfield Inlet	146	316	170	116.4
Clyde River	40	498	458	1145.0
Coppermine	230	945	715	310.9
Coral Harbour	117	537	420	359.0
Gjoa Haven	98	742	644	657.1
Grise Fiord	70	52	−18	−25.7
Holman Island	98	335	237	241.8
Igloolik	133	937	804	604.5
Iqaluit (Frobisher Bay)	512	3,126	2,614	510.5
Inuvik	1,248	2,773	1,525	122.2
Lake Harbour	90	350	260	288.9
Nanisivik	0	319	n/a	n/a
Pangnirtung	114	1,087	973	853.5
Pelly Bay	94	345	251	267.0
Pond Inlet	53	909	856	1615.1
Rankin Inlet	586	1,440	854	145.7
Repulse Bay	116	465	349	300.9
Resolute Bay	153	164	11	7.2
Sachs Harbour	76	145	69	90.8
Spence Bay	124	564	440	354.8
Tuktoyaktuk	409	970	561	137.2
Whale Cove	125	230	105	84.0

SOURCE: Krolewski 1973: 20 and Northwest Territories 1990: 4.

is not new. Over the past thirty years, many such communities have doubled their size and some in the Arctic have exceeded this level of increase. From 1961 to 1989, Pond Inlet's population increased by over 16 times while Iqaluit's size increased by 5.1 times (Table 4.8). While part of this population increase was due to people moving from the land to these centres and from intercommunity movement, much of it was due to the high rate of natural increase, often over 3% per year. Over the next decade, this high rate of natural increase seems likely to continue. If it is maintained over the next 25 years and no out-migration from the Arctic occurs, then a doubling of population will occur within 25 years. While estimating population change over 25 years is speculative, substantial growth is likely to continue to the year 2000. What is less certain, however, is which communities will grow. The most likely candidates are those regional centres that house a number of government agencies providing a public service base attracting people and offering a modest amount of wage employment. These centres include Inuvik in the Beaufort Sea area, Cambridge Bay in the Central Arctic area, Iqaluit in Baffin Island area, and Rankin Inlet in the Keewatin area.

A fourth characteristic is that the urban functions are controlled by cities either on the southern edge of the North or outside the region. For example the six largest regional centres are Chicoutimi-Jonquière (158,468), Thunder Bay (112,272), Sault Ste Marie (80,905), Prince George (67,621), Timmins (46,657), and Prince Albert (40,841), while the large metropolitan cities controlling most of the trade found in the northern hinterland are Montreal, Vancouver, Edmonton, and Winnipeg. They have extended their trade areas into the North and dominate retail and wholesale trade. Similarly, when resource projects do take place in the North, most of the benefits flow to these large centres. For example, the largest manufacturing jobs associated with the Norman Wells Oil Expansion and Pipeline Project were the central processing facility and the steel piping. The central processing plant was constructed in Edmonton, while the pipes were made in Edmonton and Regina.

Labour Force

The workforce of a region represents those persons who could produce goods and services if there was a demand for their labour and if they desired to participate in such activities. Often it is equated to the concept of the potential labour force, defined as those persons over the age of 14 and under the age of 65. One measure of the North's labour force is the dependency ratio. This ratio compares those people in the 'dependent' ages (under 15 and over 64 years) to those in the 'economically productive' ages (15 to 64 years). The dependency ratio provides a general indication of the economic burden that the productive portion of a population must carry. The implication is that the higher the ratio, the greater the burden on those in the economically productive ages. The dependency ratio for Canada in 1986 was 47, meaning that there were 47 persons either under 15 years of age or over 64 years of age for every 100 persons in the working age group. The dependency ratio for northern Canada is above the national average. Of the 29 northern census divisions, 24 had ratios above the national average and five (Saguenay 40, Yukon 40.4, Fraser-Fort George 45.1, Chicoutimi 45.7, and Stikine 46.2) had dependency ratios below the Canadian figure, indicating that a much greater proportion of the northern population lies in the dependent age group than for the nation as a whole, and that five census divisions have dependency ratios similar to those found in Third World countries.

Vignette 4.5   Urban Primacy and the Rank-Size Rule

Geographers have long observed that in many western countries there was an apparent 'order' to the size of cities but not in developing countries. That is, in developed countries the second largest city was often close to half the size of the largest centre; the third largest was one-third the size of the largest, and so on. This relationship is called the rank-size rule and is expressed as:

$$P_r = P_l/R,$$

where

P_r is the population of the rank of a particular city,
P_l is the population of the largest city, and
R is the rank of that particular city in the region or country.

In 1961, Berry attempted to link the notion of the rank-size rule to levels of development, In other words, as a country proceeds through the so-called development phases (which may be described as early industrial, late industrial, and post-industrial phases), the urban system changes from having one extremely large city in the developed area of the country and many small centres in its hinterland to one with a dominant city and several regional cities of varying size, all surrounded by towns and villages. Urban primacy was therefore supposed to occur when a country was in an early stage of economic development. Berry's findings were later challenged by El-Shakhs (1972). While scholars still argue about the meaning of such urban patterns, primacy is commonly found in developing countries or regions like northern Canada.

SOURCES: Berry 1961, El-Shakhs 1972, and Johnston, Gregory, and Smith 1986.

While the North is described as having a resource economy, most jobs are found in the public sector. Employment figures for the Northwest Territories indicate that government employment at the federal, territorial, and municipal levels forms around 44% of all employment (Northwest Territories 1990: 13). Jobs produced by the resource industry, while often high-paying, form a smaller proportion of the total labour force. A recent Yukon labour market activity survey demonstrates this point. Based on the working experience of Yukoners between June 1987 and May 1988, only 6% of the almost 27,000 jobs reported were classified as primary jobs, involving fishing, trapping, forestry, logging, and mining, while those in the managerial and professional group (many of whom are public employees) formed the largest group at 26% (Yukon 1990a). The resource industry generates a number of jobs in other sectors of the economy and, from this perspective, this industry plays an important role in job creation through indirect job creation (Vignette 4.6).

Another characteristic of the northern work force is that many jobs are seasonal. This is particularly true in the construction and tourist sectors where much of the employment occurs during the short summer period. The Yukon work force provides examples of other characteristics. In 1989, for instance, job tenure reveals that men had held their jobs longer than women, non-Natives longer than Natives, and married persons longer than unattached persons. Education was an important factor for all employees. On average post-secondary graduates held their jobs the longest (Yukon 1990b).

The last characteristic is that the percentage of persons 15 years of age and older who are in the workforce (labour force participation rate) is much lower among Natives than among non-Natives. In the Northwest Territories, for instance, in 1986 the labour force participation rate for Natives was 55% and 87% for non-Natives; Native women had the lowest rates, at 48%, compared to non-Native women at 79%. Similar figures are found in other areas of the North, reflecting a lower participation rate by Natives in general and Native women in particular.

Vignette 4.6   Mining Employment

Traditionally, mining has been the largest private sector employer in the Territories. Over the years, mining employment has remained relatively stable at about 11 per cent of the labour force, or about five times the Canadian average, while accounting for up to 25 per cent of wages and salaries. In addition, mining creates a significant number of indirect jobs in transportation, electricity, construction, government, and other service sectors in the territories, as well as in the rest of Canada. In much of the North, particularly in the Arctic, mineral exploration, mining, and associated service activities are the only practical alternative to employment in the public sector.

SOURCE: Verleun and Mackenzie, 1988: 17–18.

Summary

The population of the Canadian North has grown from around 60,000 persons around the time of Confederation to 1.5 million by 1981. Virtually all of this population increase took place in the Subarctic. This increase can be attributed to the development of northern resources and the influx of southerners to fill those jobs. During the 1980s, however, there was a halt to population growth and the population of the North has apparently stabilized around 1.5 million. The main reason for this decade of 'no growth' is due to the economic down-turn in the resource industry. Another reason is related to the trend toward more and more capital intensive resource projects which require few workers. Lastly, resource industries are making more use of southern workers and flying them to remote workplaces rather than building resource towns.

From 1981 to 1986, the large out-migration recorded by the last two censuses did not cause a large drop in total population because the high rate of natural increase softened its impact. In 1986, for instance, the rate of natural increase in the North was 1.3% while it was only 0.8% in Canada.

The nature of the northern economy and the physical environment have encouraged people to live in urban places. The urban system found in the North is strongly controlled by the major metropolitan cities found in southern Canada. Most of the urban places in the North are small centres, often with fewer than 1000 persons. The few larger towns and cities in the North tend to be located along the transition zone with southern Canada.

The demographic characteristics of the northern population reflect the large number of younger people. In northern parts of Ontario and Quebec, however, the population tends to be older. Similarly, much of the North has a high dependency ratio, suggesting that many people are not in the 'economically productive' age.

The North has undergone a rapid population increase. Over the past decade, however, its population has stabilized. In the coming years, the key to population change lies in the resource industry. The trend toward more capital intensive industrial projects suggests fewer workers will be required and that may translate into a smaller northern population.

Selected Readings

Choinière, Robert and Norbert Robitaille, 1987. 'The Fertility of the Inuit of Northern Quebec: A Half-Century of Fluctuations', *Acta Borealia* 1/2:53–64.

Maslove, Allan M. and David C. Hawkes, 1990. *Canada's North, A Profile*. 1986 Census of Canada (Catalogue 98–122). Ottawa: Minister of Supply and Services.

Robitaille, Norbert and Robert Choinière, 1987. 'The Inuit Population of Canada: Present Situation, Future Trends', *Acta Borealia* 1/2:25–36.

Wonders, William C., 1987. 'The Changing Role and Significance of Native Peoples in Canada's Northwest Territories', *Polar Record* 23/147:661–71.

Chapter 5

RESOURCE DEVELOPMENT

The northern economy has always been based on the exploitation of its natural wealth. In pre-contact times, the major resource was wildlife; after contact with Europeans, it was fur-bearing animals; and now it is a wider mix of resources, including energy, forests, and minerals. During the 1980s, the annual value of resource production from the North was in the billions of dollars. In 1989, the Northwest Territories accounted for $1.2 billion worth of resource exports (GNWT 1990: 109). A conservative estimate of the total value of all northern resource production in 1990 would be $30 billion.

In the 1950s, the resource economy offered a fresh economic approach for northern Canada. Increased demand for energy and raw materials, and federal financial support for development not only created a new economic environment in the North but also expanded the national, provincial, and territorial economies. The North soon became an important staple hinterland serving Canada, the United States, and other industrialized nations. The new economic landscape consists of mines, oil wells, pulp mills, and hydroelectric power stations, all connected to southern markets by a modern transportation network. The construction of these industrial projects and the building of a complementary infrastructure consisting of towns, roads, and a wide range of public facilities generated a demand for goods and equipment produced in southern factories; created job opportunities in both northern and southern Canada; and increased the volume of Canadian exports, thereby improving Canada's balance of trade. Both levels of government have benefited from increased tax revenues generated by the resource industries and the associated service industries, as well as from their employees' personal income taxes.

The Demand for Northern Resources

At first, the increased demand for resources came mainly from the United States. In the early post-World War II years, American industrialists were unable to satisfy their need for raw materials from within their borders, and they began to look at securing these resources from foreign countries. Canada's North was an attractive area for three reasons. It contained the desired resources, relatively close to factories in the United States; Canada was perceived as a politically stable state and therefore a safe place for American investments; and Canadian resources were viewed by American firms and governments as falling within a North American trading bloc. In 1947, American iron and steel interests announced plans for the first large-scale, privately funded resource project in northern Canada. This plan called for the exploitation of the vast iron deposits in northern Quebec and Labrador and the building of a rail-sea transportation system capable of supplying iron ore to American steel plants. By the early 1950s, this massive iron mining and transportation system was completed. In more recent years, other foreign countries have turned to the resources of northern Canada. Japanese companies, for example, have leased enormous tracts of timber in northern Alberta for their pulp and paper plants. Canadian resource companies have also been drawn to the North, particularly to smaller projects requiring less capital and with a relatively short pay-off. The high risk of large-scale projects is real. The closure of the Cyprus Anvil mine at Faro, Yukon and the bankruptcy of the leading oil exploration firm in the Beaufort Sea, Dome Petroleum Company, represent the consequences of such risk. Offsetting the possibilities of economic failure, and drawing companies to the North, are the prospects of large profits.

Accessibility and Resource Development

In the past, resource development in the northern frontier was often stalled by the absence of a modern transportation system. The cost of building the necessary transportation link to the south was usually beyond the financial capacity of the developer. Over the years this handicap has been reduced, but the basic problem remains for much of the North and certainly for the Arctic.

The high cost of construction projects in the North is due to four main factors. These are the physical terrain, particularly the Canadian Shield and the presence of muskeg and permafrost; the short construction season; the need to assemble men and equipment in remote areas; and the long distance from southern supplies. Because of the short construction period in the summer, and the lower labour productivity associated with winter construction, costs are much higher in the North than in the rest of the nation. If the workforce, equipment, and supplies must be assembled in the south and then transported to a remote northern location, the costs are extraordinarily high. The presence of permafrost at the construction site calls for special–and costly–construction methods. For the same reasons, maintenance of rail lines, pipelines, and highways is much more costly than in southern Canada.

Because of these high costs, resource development which has minimal transportation needs often occurs first. Historically, the more valuable resources are exploited first, e.g., gold and silver. When refined, these highly valuable minerals can be shipped to market by small aircraft. In such cases, there is no need for an expensive all-season highway or railway. On the other hand, base metals like lead and zinc remain bulky products even after milling the ore. They require rail, road, or sea transportation. The degree to which various mineral deposits are deemed exploitable is largely a function of the value of that commodity per unit of weight, that is, the higher the per-unit value of the mineral, the greater the distance it can be shipped.

Both provincial and federal governments have tried to overcome this distance barrier to resource development by building modern transportation routes into the North. In the 1920s, the British Columbia government tried to stimulate resource development in its northern territory by building the Pacific Great Eastern Railway (PGE). In 1952, some thirty years later, the PGE was extended to Prince George and North Vancouver, thereby providing an effective link from the main port on the west coast to the regional capital of the northern interior of British Columbia. At the same time, the PGE reached further north to Fort St John and further east to the Dawson Creek and the Peace River country. At Dawson Creek, the PGE established a link with the Northern Alberta Railway. During the 1960s, resource development of northern British Columbia began in earnest and the PGE both assisted and benefited from the resource boom in forestry and energy. This railway was renamed British Columbia Railroad in 1972 and BC Rail in 1984.

In spite of the efforts of individual provincial governments, the Canadian North remained a remote and inaccessible place. For resource companies, this remoteness meant that known resources in the Arctic and Subarctic were not economical to develop. From their perspective, the cost of building northern roads or railways to ore bodies was too great and, if the North were to be opened up by the private sector, the federal and provincial governments had to provide suitable access routes.

Under the terms of Confederation, the federal government had no responsibility for building roads in the provinces and, until the 1950s, there was little need for roads in the territories. But circumstances changed and new policies were needed. By the late 1950s, the old laissez-faire policy was abandoned, largely because the federal government saw northern development as a way to strengthen the nation. As well, because it had initiated a policy to 'modernize' the North, Ottawa hoped resource development would provide jobs to Native northerners who were now living in settlements. In 1958, there was a major shift in federal government policy. The newly elected Conservative government sought to increase the rate of resource development in the North by using public funds to extend the southern transportation system into the North. In its Roads to Resources program, the federal government agreed to fund half the cost for building roads in northern areas of the provinces, leaving the provinces responsible for the remainder. A similar program was available in the territories, but in this case the federal government covered all the costs. Under these two programs, roads were built to help private companies reach world markets. One example is the road built to connect the asbestos mine at Cassiar in northern British Columbia with the sea port of Stewart. Other highways were constructed to provide major northern centres with a land connection to southern Canada. For example, the Mackenzie Highway from Edmonton to Hay River was extended to Yellowknife. In all, from 1959 to 1970, over 6,000 kilometres of new roads were built at a cost of $145 million (Gilchrist 1988: 1877).

Vignette 5.1   The Hudson Bay Railway

The importance of modern transportation to resource development is evident in the few instances of rail lines built into the Subarctic. The Hudson Bay Railway provides one such example. Completed in 1929, it was designed to provide an alternative route to European markets for western grain. Prairie farmers believed that the shorter Hudson Bay rail route to their major European customers would reduce their total shipping costs. Unfortunately the new railway never became an important export route for grain. Instead, the Hudson Bay Railway was instrumental in unlocking the mineral, forest, hydro, and fishing resources of northern Manitoba. Unlike the original dream of the Prairie farmers, however, the Hudson Bay Railway did not serve to export grain through the Port of Churchill but rather to bring resource products south into the markets of southern Canada and the United States.

Social and Political Implications

The economic development of the North was closely followed by social and political changes. The federal, provincial, and territorial governments now provide basic education and health services to all communities in the North. The cost of building this education and health infrastructure has been considerable and its annual operating cost high. In the territories and the more remote areas of the provinces, the local tax base is far too small to pay for operating these institutions and most operating funds come from federal and provincial sources. Besides the transfer of public funds into the North, political power has been shifted from Ottawa to the two territories. This devolution of powers has not run its full circle to provincial status but it has come a long way in a short period of time. No such transfer of powers has occurred in the provinces' northern areas.

These developments have exposed northern Natives to a different way of life, one that they are encouraged to join. Much of this encouragement takes the form of education and job training. So far, the participation of Native adults in the work force is low. Though it has increased over the past half-century the percentage of Native adults employed is far less than the percentage of the non-Native population. The alternative source of income for Native northerners is trapping. With low fur prices and high costs of outfitting a trapper, few northern Native Canadians are able to satisfy all their needs from the land. The Native economy now involves periodic wage employment and transfer payments. Many families cannot obtain sufficient income from the combination of trapping and wage employment, and have become dependent on welfare payments and public housing programs.

From the Native perspective, the cultural cost of participating in this wage economy is high–too high in the eyes of some. Substantial social changes are required, such as living in a resource town, working in a non-Native environment, and participating in a lifestyle practised by fellow workers. Few Natives have made this change. Most still live in Native communities where there are few jobs but where the social environment is a Native one. The complex cultural and economic issues affecting Native Canadians are more fully discussed in subsequent chapters.

The Nature of Resource Hinterlands

The Canadian North is but one of many resource hinterlands. All hinterlands have a number of common characteristics.

(1) World demand for primary resources and energy determines the course of hinterland development.
(2) Multinational corporations, with their capital, management skills, and technical knowledge, are the leading force in resource development.
(3) The global demand for raw materials and energy is cyclical, following the global business cycle. These cycles are more pronounced in resource hinterlands, leading to a 'boom-and-bust' economy.
(4) Resource hinterlands in other countries often compete against each other, driving the price of primary products down.

(5) Primary resource exploitation in resource hinterlands is associated with severe economic leakage: most of the economic benefits generated by resource projects find their way to other, more developed regions.

(6) Most resource projects characteristically require a small but trained labour force. The oil and gas industry, for instance, tends to recruit experienced workers from other oil patches.

(7) Large construction projects employ large numbers of workers for short periods of time.

The northern hinterland of Canada reflects these characteristics. External demand for resources and energy has shaped the North's economy. This production is export-oriented, responding to the needs of industry found in southern Canada, the United States, and other industrial countries. While the Subarctic is much more fully integrated into this world economic system than is the Arctic, the pattern of development is the same: resource production from both regions serves world markets. In some cases, minerals are shipped directly to foreign countries where they are used in the production of finished products. For example, coal from the Tumbler Ridge mine in northern British Columbia goes directly to Japan where it helps fuel the Japanese steel industry. In other cases, domestic markets are served first. For instance, energy shipped by pipeline or electrical power lines serves provincial markets first and then American ones.

Northern development is largely in the hands of multinational resource companies. These firms have invested billions of dollars in resource projects and, along with public funds, have created northern towns, provided many jobs, and built transportation systems. All these efforts have led to the development of northern Canada and to the profitability of these firms. In northern Saskatchewan, for example, the main resources are uranium and timber. Weyerhaeuser Canada, an American-owned firm, controls most of the commercial forest in northern Saskatchewan through its timber licence. Ownership of the uranium mines is more complicated and involves two former Crown corporations, Eldorado Nuclear Ltd and Saskatchewan Mining and Development Corporation which amalgamated into Cameco (Canadian Mining & Energy Corporation). In the 1970s, the three open-pit uranium mines (Rabbit Lake, Cluff Lake, and Key Lake) were owned and operated by Gulf Minerals, an American-controlled company, Amok Ltd, a French firm, and Uranerz Exploration and Mining Company, a West German firm. In 1991, both Rabbit Lake (now called Collins Bay) and Key Lake are operated by Cameco through joint ventures with several multinational companies. Cameco itself is owned by the government of Saskatchewan (61.5%) and the federal government (38.5%). Amok continues to operate the Cluff Lake mine.

The northern economy, like that of other hinterlands, is vulnerable to

Figure 5.1
Yukon Gross Domestic Product, 1977–1987

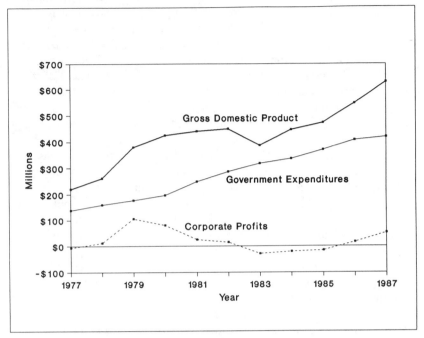

SOURCE: Yukon Bureau of Statistics, 1990.

the 'boom-and-bust' resource cycle, characterized by a rapid increase in economic activity followed by a sudden drop in economic activity. Unfortunately, the occurrence of the 'sudden decline' is not predictable and most people are caught off guard. The uncertainty caused by variations in global demand makes it difficult for workers and their families to make a long-term commitment to the region. Many areas of the North have witnessed the opening and closing of a mine; its impact on a community is devastating and often results in the abandonment of the settlement. In 1981 Uranium City had a population of nearly 2500. When the Eldorado uranium mine was closed the town lost its economic base and, within a year, its population plummeted to less than 500. By 1991, only a few Native families remained. In the early 1980s, another example of the impact of a mine closure on a small regional economy took place in Yukon. In 1983 the closing of the lead-zinc mine at Faro depressed the entire Yukon economy, reduced government revenues, and triggered an out-migration. Yukon gross domestic product dropped by over 10% from 1982 to 1983 (Figure 5.1).

Multinational corporations are global operations and often have resource projects in various hinterlands. Inco, for example, has a nickel mine in Indonesia as well as at Sudbury, Ontario and Thompson, Manitoba. Multinational corporations may also keep the price of their raw materials low by financing a number of suppliers of a particular resource. This approach may involve long-term purchase agreements, leaving the actual production to another corporation. The Tumbler Ridge coal mine in northern British Columbia was financed through a coal-purchase agreement with Japanese steel firms. When the world price for coal dropped, the Japanese firms renegotiated the price of Tumbler Ridge coal, demanding that it reflect world price. The Tumbler Ridge coal operation is not viable at current world prices and may be forced to curtail its operations.

Construction of resource projects often costs billions of dollars. They require expensive equipment which is produced outside the region. A billion-dollar construction project might see virtually all of the equipment and supplies purchased in southern Canada or in foreign countries. In the construction of the Norman Wells Oil Expansion and Pipeline Project between 1982 and 1985, the major expenditures for supplies and equipment occurred in southern Canada. Esso Resources Canada Ltd let the contract for building the complex central oil-processing facility to a firm in Edmonton, while Interprovincial Pipe-line (NW) Ltd purchased pipe for the pipeline from steel plants in Edmonton and Regina. This geographic pattern of purchases, common to all resource projects undertaken in the northern hinterland of Canada, shifts the economic benefits to southern Canada.

Resource industries vary in the number of people employed. The forest industry is the principal employer, followed by the mining industry. Hydroelectric and pipeline operations employ very few workers. The large hydroelectric projects in British Columbia, Manitoba, Quebec, and Newfoundland (Labrador) have produced few permanent jobs in the North with the exception of those projects associated with the smelting of copper at Flin Flon and bauxite at Arvida and Kitimat where aluminum is produced.

Resource Differences in the Arctic and Subarctic

The geographic distribution of resources across the North varies by climatic and geological regions. The Subarctic contains vast forests while the Canadian Shield holds rich mineral deposits. Added to these natural factors are economic factors such as accessibility to markets. Across these vast lands, therefore, there are significant differences in resource potential, accessibility to export markets, and costs of development. Equally important is the question of sustainable resource development. Such development is only possible in areas where there are renewable resources.

The resource bases of the Arctic and the Subarctic differ significantly.

The Subarctic has a number of advantages in each area: its resource base is much broader and includes major energy, mineral, timber, and water resources; its geographic location makes it more accessible to the American market; and its natural environment presents less of a barrier to industrial projects because of a milder climate, a longer navigation season, and less land affected by permafrost. All of these advantages translate into lower costs of resource development in the Subarctic than in the Arctic. In the 1930s, mineral deposits were being exploited, including mines at Goldfields, Saskatchewan, Keno City, Yukon, Kirkland Lake, Ontario, Sherridon, Manitoba, and Yellowknife, Northwest Territories. Much earlier, timber was cut across the southern edge of the boreal forest in Quebec and Ontario, and rivers flowing southward from the Canadian Shield were used to bring the logs to saw mills and pulp mills in southern Ontario and Quebec. By the turn of the century, logging had become a key primary economic activity in western Canada. Centres like Prince Albert, Saskatchewan owed much of their early economic growth to the timber industry.

In the post-World War II decades, resource development took a new form: mega-projects. All these billion dollar schemes were built in the Subarctic. Mega-projects involved billions of dollars of investment and matching public funds were spent on transportation and settlement infrastructure. Initially, the driving force behind these enormous industrial undertakings was American industry. No longer able to supply themselves with raw materials and minerals from within the United States, American corporate giants turned to the rich primary resources of northern Canada. By the late 1950s, a resource boom was underway in the Subarctic. This prompted pro-developers like Richard Rohmer to refer to the Subarctic as the 'Green North'. By the 1970s, massive Canadian and offshore investments also took place. By the 1980s, the Subarctic was no longer a frontier, but an integral part of the global economy.

In contrast, little resource development had occurred in the Arctic. By the late 1950s, the only mine in the Arctic was the North Rankin nickel mine, a 'marginal and very small-scale operation', operating from 1957 to 1962 (Rea 1968: 148). Three other mines, Asbestos Hill, Nanisivik, and Polaris, have since come into production. Asbestos Hill mine, located along the shores of arctic Quebec, closed in 1983. Nanisivik is located on the northern tip of Baffin Island, and Polaris on Little Cornwallis Island in the Arctic Archipelago. Sea access was one factor that made these mines commercially viable. Large oil and gas deposits have been discovered in the Beaufort Sea and in the Arctic Archipelago. Token shipments of crude oil were sent from the Bent Horn field in the Arctic Archipelago to Montreal in 1985 and 1986, and from the Amauligak field in the Beaufort Sea to Japan in 1986. The more accessible Beaufort petroleum and gas deposits are first in line for commercial exploitation. This energy development is expected to be the most costly mega-project ever built in the Canadian North.

The Arctic has yet to be fully drawn into the global economic system. The delay in utilizing arctic resources is due to the high cost associated with exploration, construction, and transportation. Both existing arctic mines, Polaris and Nanisvik, are able to ship their low value product (lead and zinc) by sea to European markets. They store their production for about 10 months and then ship it to European markets during the short navigation season. The Lancaster Sound sea route to the Polaris mine site on Little Cornwallis Island may be filled with floating ice at any time during the short navigation period. Both companies ship their freight and concentrated ore in the MV *Arctic*, which has a double steel-reinforced hull, advanced radar, sonar, and satellite navigation systems. Neither mine site has a town as all workers are flown to the work site but live in other communities in the rest of Canada (and in one or two cases, in other countries). When these mines close, the work site at Polaris is expected to be abandoned, while Nanisivik may be converted into a military base.

High Wages and Living Costs

The resource industry requires a skilled labour force to operate its hydro-electric stations, mines, oil refineries, pulp plants, and pipelines. It has recruited managerial, professional, and trades employees from southern Canada and other industrial nations. Foreign-owned firms often hire managerial and professional workers from their country. Employees are attracted to the North by high wages, which offset the loss of amenities found in southern Canada. Take-home pay by miners is particularly high because they often work long shifts, receive bonuses, and are taxed at a lower federal rate. In analyzing 1961 census data, Rea (1968: 182–4) reported that in 1961 salaries and wages in underground gold mines at Yellowknife were, on average, 45% higher than those in Ontario. Since then, incomes in resource industries have remained high, pushing the average income for all workers in the two territories above all provinces except Alberta (Table 5.1).

High incomes are partly offset by the cost of living in the North. Consumer goods, whether food or clothing, are imported from the south and northern prices include transportation costs and extra handling expenses. In addition, most northern retail stores are small and cannot take advantage of economies of scale. Living costs are highest in the more remote areas of the North. One measure of these differences is illustrated by the isolated-post index for federal government personnel employed in the Northwest Territories while another is the Government of the Northwest Territories' food-cost index (Table 5.2). Both indexes indicate that the highest living and food costs occur at Spence Bay, an isolated community beyond the Arctic Circle. All foodstuffs, clothing, and other goods must be imported from the south by air. Prices at Spence Bay

Table 5.1
Personal Income Tax Returns: Average Income, 1986

Rank	Province/Territory	Average Income
1	Alberta	$19,750
2	Northwest Territories	19,525
3	Yukon	19,480
4	British Columbia	19,102
5	Ontario	18,955
6	Saskatchewan	17,134
7	Quebec	16,673
8	Nova Scotia	15,790
9	Manitoba	15,425
10	New Brunswick	14,462
11	Prince Edward Island	13,603
12	Newfoundland	13,221
	Canada	17,896

SOURCE: Government of the Northwest Territories, Bureau of Statistics 1987: 7.

Table 5.2
Prices in the Northwest Territories

Place	Federal Isolated-Post* Living Cost Differentials, 1988–90		GNWT Food Price Index 1987**
	Price Index	Base City	Food Index
Hay River	120–124	Edmonton	89
Yellowknife	125–129	Edmonton	100
Fort Simpson	135–139	Edmonton	101
Inuvik	150–154	Edmonton	124
Iqaluit	155–159	Montreal	151
Tuktoyaktuk	160–164	Edmonton	122
Rankin Inlet	160–164	Winnipeg	146
Grise Fiord	170–174	Montreal	178
Pond Inlet	170–174	Montreal	179
Coral Harbour	180–184	Winnipeg	168
Sachs Harbour	190–194	Edmonton	170
Pelly Bay	195–204	Winnipeg	192
Spence Bay	200–204	Edmonton	196

*Isolated-post indexes are prepared by Prices Division, Statistics Canada. They are used to determine allowance levels for federal personnel employed at isolated locations. As of September 1989, four city indexes were established based on the Canada average index of 100. These cities are the main suppliers to the North and their indexes were Montreal 102, Winnipeg 95, Edmonton 95, and Vancouver 99.

**Food-price indexes are prepared by the Bureau of Statistics of the Government of the Northwest Territories. Food price information was gathered from each community in 1987. Expenditure data used for weighting purposes is based on Statistics Canada's 1982 survey of Yellowknife family expenditures. Yellowknife's food costs are set at 100.

SOURCE: Government of the Northwest Territories 1990: 71–2 and 130–1.

reflect both high air shipping rates and long distances from southern suppliers. A recent report (GNWT 1986: 2) summed up the situation as: 'while average personal income, at $25,000 for the NWT as a whole, was about 4% greater than the Canadian average in 1984, the cost of living in the NWT is approximately 44% higher than it is in southern Canada.'

The Public Role in Resource Development

Until the 1950s, the federal government was not a conspicuous factor in northern development. The two territories were left to the missionaries, fur traders, and Native people. A similar pattern took place in the northern areas of the provinces. Here, the federal government's role was restricted because provincial governments were responsible for the construction of transportation routes and for the management and taxation of natural resources situated within their respective jurisdictions. Each province with northern lands was responsible for unlocking its own resources. Little was done, however, because the demand for northern resources was still low and both provincial and federal governments were loath to spend scarce public funds in the North. A measure of the parsimonious attitude to northern territories is revealed by the fact that in 1946, not a single school had been built by any government in the Northwest Territories except for that in the mining town of Yellowknife (Rea 1968: 52).

In the early 1950s, federal and provincial governments took a more active role in northern development. Both governments recognized that the high cost of resource development necessitated a partnership approach. At first, the public sector supported private ventures by providing access to remote sites, and funds for the social infrastructure of single-industry towns. The federal government developed a number of new initiatives, among them the Roads to Resources program. Crown corporations were another public instrument. Provincial Crown corporations focused their attention on water resources while the federal government took an active role in the petroleum industry. In the post-World War II period, a number of provincial governments created public power companies. British Columbia, Manitoba, and Quebec built large hydroelectric projects to produce electricity for the local provincial market. Excess power is sold to American utility companies in adjacent parts of the United States. Since the Free Trade Agreement with the United States in 1989, energy and other resources have full access to American markets. The flow of raw materials and power to the United States is likely to increase in the 1990s.

Two factors, northern development and foreign ownership of much of the petroleum industry, caused the federal government to form Petro-Canada, a federal Crown corporation. It was established in 1975 to 'Canadianize' the industry and to explore and develop frontier oil and gas deposits. Petro-Canada's initial expansion was accomplished by acquiring the Canadian assets of four foreign-owned multinational companies. By 1991, Petro-Canada had become one of the largest petroleum corporations in Canada with assets over $10 billion.

The federal government also promoted resource-development by providing generous grants and incentives to private resource development companies. The most expensive subsidy program was the Petroleum Incentive Program (PIP) which poured billions of dollars into frontier energy exploration (Vignette 5.2). PIP allowed a form of super-depletion allowance which could form up to 80% of the costs of drilling for Canadian companies and 25% for foreign companies. This level of public support, while it accelerated the exploration of the Beaufort Sea oil and gas deposits, slowed petroleum exploration in the provinces. While PIP did lead to major oil and gas discoveries in the Beaufort Sea, these proven reserves are still far too expensive to develop at 1991 oil prices.

While most public subsidies have gone to large private companies, northern co-operatives have until recently received much less support from the federal government and the Government of the Northwest Territories. Co-operatives play an important role in fostering community development and Native involvement in the commercial affairs of their communities. A federation of all Northwest Territories community co-operatives was created in 1983 when the Arctic Co-operatives Limited (ACL) was formed. Arctic Co-operatives Limited provides marketing and purchasing services to retail co-op stores in over 30 communities in the Northwest Territories. It also assists local staff development and training. In 1979, the ACL received $2 million for a management training program from the federal Special ARDA program. The objective of this program was to train six Native persons as co-op managers and ten as department managers (Abele 1989: 84). Eight other employees were to have increased responsibilities. From 1979 to 1983, 49 trainees enrolled in the program. At the end of the four-year training period, 20 were employed as managers or department managers and nine had taken management positions in other firms, while 15 did not complete the program. Job training programs, while expensive, accelerate the process of placing Native employees in managerial positions and thereby reduce the need to recruit southern workers. The funding represents an investment of approximately $10,000 per year for four years for each trainee.

Federal Transfer and Deficit Payments

The North, with few people and little industry, has a small tax base. Given its enormous size and physical character, the cost of providing public services in the North is much higher than in southern Canada. Tax revenues collected in southern Canada help pay for the delivery of services in the North. These taxes are collected by the federal government and

---

Vignette 5.2   Trudeau's Answer to Multinational Foreign Oil Companies

In the early 1970s, the Trudeau government created the Foreign Investment Review Act (FIRA) (1974) and established the Crown corporation, Petro-Canada. These political actions were designed to break the dominance of the multinational oil and gas companies in Canada. Petro-Canada, like the foreign-owned multinational companies, has the financial strength, technology, and long-term perspective to undertake large, expensive projects like oil sands development and arctic exploration. In 1980, the government announced the National Energy Program, which increased the federal taxation of petroleum revenues and took away the tax write-offs that had encouraged exploratory drilling in the 1970s. In place of these tax write-offs, the government created the Petroleum Incentive Program (PIP), which paid grants to Canadian-owned firms undertaking oil exploration. These grants were based on the level of Canadian ownership and the location of the wells. The largest grants were awarded to companies with at least 75% Canadian ownership, drilling for petroleum in frontier areas like the Arctic.

---

then redistributed to the two territories and have-not provinces. Unlike the provinces, the two territorial governments are not eligible for federal assistance through the equalization program. As a substitute, the federal government has developed special funding arrangements with the two territorial governments. While the per capita federal payments to the territories are much greater than those made to have-not provinces, the taxing powers of the provinces include natural resources; in the two territories, the federal government collects taxes from resource industries. Since 1988 the federal and Northwest Territories governments have been discussing the transfer of oil and gas royalties to the territorial governments but so far have reached no agreement. The main stumbling block is federal reluctance to surrender off-shore oil and gas fields to the territories. Since Canadians have made a commitment to provide similar education, health, and social services in all parts of Canada, the two territories will continue to require financial assistance from the federal government.

While most Canadians accept the principle of equalization payments or deficit grants, one group of economists challenges the idea. They argue that such payments create a transfer-dependency mentality, preventing have-not provinces and territories from making the hard economic decisions to reform their local economy (Courchene 1986: 25–62). Transfer dependency arises when a province or territory cannot raise sufficient revenues to cover the level of services expected by Canadians. The ultimate solution, in these economists' view, involves the lowering of real wages which would then attract industry or, failing that, shifting surplus labour from slow growing regions to fast growing ones (Matthews 1983: 60). The flaw in this argument is that such measures are likely to widen the gap between the rich areas of Canada and the North, causing more economic and social distress in the North.

Originally, deficit payments were determined by the difference between the total revenues raised by each territory and its total expenditures (Robertson 1985, and Stabler 1987). Ottawa recognized that the territorial governments had no incentive to increase their tax rates (which are well below those in the provinces) or to find more efficient ways to deliver services to territorial residents. In 1989, the federal government changed the 1985 Formula Financing for determining deficit payments. The new formula limits the annual increase in territorial deficit payments; in turn, the territorial governments are forced to priorize the ever-growing demand for more services, and to raise more revenue from territorial residents. These changes should lead to more fiscally responsible governments in the territories, and encourage more efficient governmental operations; failure to do so would force the government of the Northwest Territories to raise its taxes, which are much lower than those of the provinces.

Even with the new deficit formula, federal payments to both territories are expected to rise for the foreseeable future. With the much higher natural rate of population increase in the Northwest Territories, deficit payments to the Government of the Northwest Territories are expected to increase more rapidly than those for the Government of Yukon. Such funding from the federal government is necessary, and will have to continue for a long time for at least four reasons. First, the territorial governments are limited in their ability to raise taxes: unlike the provinces they do not have the authority to collect taxes from firms exploiting minerals and energy. These taxes go to the federal government. Second, the territories' small population and business community limit the tax base. The territorial governments cannot generate the necessary revenue to allow them to operate at their current expenditure levels. Third, the large number of very small communities in the North cannot achieve the same level of economies of scale as do larger centres, and have higher per capita costs for delivering a variety of public services. Lastly, the vast distances between northern centres incur transportation costs that add substantially to northern governments' total cost of delivering public services.

The Political Geography of the North

The North's nine provinces and two territories each try to capture the benefits of resource development within their borders by various programs. Perhaps the most successful program has been the construction of transportation routes from major provincial cities to the northern hinterland. Other programs have been to lease large oil and gas leases and tracts of timber to multinational companies. The resulting economic landscape consists of a series of northern economies, each lying within the trade area of a southern metropolitan city. Four metropolitan cities, Montreal, Winnipeg, Edmonton, and Vancouver, have extended their trade areas beyond their provincial boundaries and into the territories. Those communities without road access to southern cities are linked to the south by regional air lines. During the short navigation period, supply ships and barges operate in the Eastern Arctic, Hudson Bay, and Mackenzie River.

In the provinces, northern residents comprise only a small proportion of the provincial population. For this reason, they have little influence over decisions taken in the provincial legislatures. Northern concerns and interests are often not recognized or are assigned a lower priority than southern concerns and issues. Accordingly, northerners living in provinces have little political voice in their respective capitals and virtually no control over their economic destiny. Examining the political situation in Northwestern Ontario, Weller (1977: 754) makes the dismal observation that 'many of the region's residents will in the future, as in the past, participate in political phenomena that could be classified as representative of either the politics of frustration or the politics of parochialism.'

Residents in the two territories have much more political influence over their government and hence over the territorial policies affecting public programs. Territorial politicians must pay attention to local (northern) issues; the same political pressure does not exist in provincial capitals. It is not surprising then that the territorial governments have a variety of public agencies and programs serving the needs of northern businesses. Provincial governments often have similar programs but they are designed to meet business conditions in southern Canada and frequently overlook the special concerns of small northern firms.

The political structure of Confederation has led to resource conflicts and resource sharing. Perhaps the best known conflict between two provinces is the dispute between Newfoundland and Quebec over the price of energy produced by the Churchill Falls hydroelectric project, the largest single-site hydroelectric project in Canada. In 1953, a consortium of five companies, mainly British, formed the British Newfoundland Corporation Ltd (Brinco) to develop Labrador's resources. Since geography dictated that electric transmission lines must cross Quebec territory to reach commercial markets, Brinco had to convince Quebec Hydro to purchase most of the electric power produced on the Hamilton (now the Churchill) River. In 1966, the Churchill Falls Corporation concluded a fixed price over 65 years with Quebec Hydro for most of the electric power produced by the proposed Churchill River hydroelectric project (Wallace 1987: 465). Brinco then formed a subsidiary, Churchill Falls Corporation, to build the dam and install the generating units. Electrical power was first transmitted to Quebec Hydro in 1971. At the time of the first sale of power to Quebec Hydro, the fixed price seemed reasonable. But as energy

Table 5.3
Economic Sectors in the Northwest Territories, 1986

Sector	Employees		Wage and Salaries	
	No.	%	$millions	%
Primary (Mining, Oil & Gas)	3,000	17	125	24
Secondary (Processing & Manufacturing)	400	2	14	3
Tertiary (Service)	14,200	80	372	71
Quaternary* (Decision-Making)	200	1	10	2
Total	17,800	100	521	100

*Estimated

Adapted from: Government of the Northwest Territories 1986: 6.

prices rose sharply after 1972, Newfoundlanders believed the fixed selling price was far too low. During the 1980s, for example, Quebec Hydro purchased electrical power from the Churchill Falls power station at about one-tenth of its market value (Wallace 1987: 465). In 1974, the Government of Newfoundland nationalized the hydro complex and then tried, without success, to renegotiate the agreement. It then appealed to the Supreme Court of Canada. In 1984, the highest court in the land found no legal basis for breaking the existing agreement.

Economic Structure of the Northern Economy

The economic structure of the North is reflected in its labour force. Labour force activities are divided into four economic areas: primary activities, secondary or processing activities, tertiary or service activities, and quaternary or decision-making activities. Given the importance of the resource industry in the North many believe that most northerners are engaged in primary activities such as logging, mining, and trapping. This assumption is not correct. The capital-intense resource industries, while generating most of the output in the North, employ relatively few workers, although their employees receive higher wages than those in the service sector. This fact is borne out in the figures for the Northwest Territories where the primary sector accounts for 17% of the employees but 24% of the wages and salaries (see Table 5.3).

Most employees work in the service sector of the economy. This is particularly true in the two territories, where there are four levels of government: federal, territorial, settlement/regional, and Native governments (band councils and regional or territorial organizations such as the Dene Nation). In the Northwest Territories, service employees form 80% of the work force with public employees making up about one-third of all service employees (Table 5.3). The large size of the public sector reflects the vast geographic extent of the North and the small but scattered population centres, all of which require basic public services. The public service sector plays a significant role in those regional economies affected by land claims settlements. In the James Bay Agreement, the Cree gained control of local government, and by 1981 approximately 35% of the Cree work force was employed in service industries (Salisbury 1986: 139).

The northern economy has a very small secondary or manufacturing sector. While most mining operations require some form of processing, this can be as simple as crushing and sorting ore or as complex as smelting ore. Few jobs are associated with the more simple processing methods. Only the bauxite smelters at Kitimat and Arvida and the copper smelters at Flin Flon and Rouyn-Noranda employ large work forces. At the copper smelters, copper ore (with less than 3% copper) is changed into a copper metal which is 99% pure copper, achieving great reduction in transportation costs. Establishing the copper and bauxite smelters in the North was encouraged by the supply of inexpensive electrical power from nearby hydroelectric generating plants. The final product (refined copper) and its manufacture into industrial and consumer products takes place in southern Canada and other industrial areas where there is ready access to skilled labour and markets. This matter of access works against the establishment of manufacturing plants in the North.

The quaternary sector, while forming the smallest part of the workforce, is often considered the most important because its workers make the decisions which set policy, manage day-to-day affairs, and determine long-term goals. This high level decision-making process is found in company headquarters and political capitals. Most Canadian private corporations have their headquarters in Toronto while foreign firms have their corporate headquarters in their national state. Comparable public decision-making takes place in the inner circles of government which are located in capital cities. Except for the two territories, the quaternary sector of the economy is not found in the North. The functions of these corporate and public headquarters include major investment and disinvestment decisions, such as opening or closing a mine. Corporate headquarters and the residences of the chief officers and board members of these companies are far from the North. Senior public officials, like their counterparts in business, live and work in a southern environment. Does the location of corporate and public headquarters in southern Canada hurt the northern economy?

Location of corporate headquarters in the North would benefit the northern economy in two ways: jobs, and decisions. The number of jobs in the quaternary sector is small but the number of indirect jobs is much larger, i.e., support staff and firms closely linked to the corporate sector. The direct and indirect employment generated by the presence of territorial governments in Whitehorse and Yellowknife is some indication of the importance of the public quaternary sector in the North. Having corporate leaders living in the North would provide them with a fuller appreciation of the North and perhaps affect their decisions regarding plant openings and closures.

The North's economic structure demonstrates four facts: (1) that the economic structure of the northern economy is heavily weighted to primary and tertiary activities; (2) that the secondary and quaternary sectors are extremely small; (3) that corporate and many public decisions about northern economic and social affairs are made outside the North; and (4) that land claims settlements may lead to a high proportion of the Native workforce entering the public sector.

While classical regional development theory suggests that a primary-based economy can evolve into a more diversified one, the geography of the North may well prevent such a transformation. The theory envisages that a staple resource (wheat, fish, timber, or whatever) triggers economic development and that this product is exported because the local market is too small to absorb the production. In the fullness of time, the region's population expands and the larger market provides new business opportunities. New firms locate in the region to satisfy the growing 'internal' demand for goods and services as well as to service the needs of the resource industry and its processing firms. During this transformation, the original resource-based section of the economy becomes less important in terms both of value of production and number of employees. Assuming that this theory is valid, the major stumbling block for the North is its small and dispersed market. If the economy of the North does not evolve into a more diversified one, what then is its future?

There is little prospect of a strong manufacturing sector. In 1986, its share of the workforce was 2%. For manufacturing firms the Northwest Territories represent a high-cost region because of distance to markets and the high cost of assembling raw materials, energy, and labour. Opportunities for processing firms are limited to refining raw materials and processing finished products for the local market. Gold mines process their ore by crushing, sorting, and melting–the final product is a gold bar which, because of its high value, can stand the transportation costs to market. Local food processing, such as a bakery, must import its raw material and, because of the small local market, operate a relatively short production line. Few such firms can survive because food products from southern Canada are mass produced (economies of scale) and marketed in the North at prices lower than those locally produced.

The service sector offers some prospects. It provides most jobs in the Northwest Territories. In 1986, the tertiary or service sector accounted for 80% of all employees and 71% of all wages and salaries (Table 5.3). The next decade will see the Northwest Territories maintain its large service sector for two primary reasons: tourism, and government. Tourism is an important element of the northern economy and generates much employment in the summer. Government, however, is the mainstay of the service sector. The figures in Table 5.3 understate the contribution of the public sector because some public agencies such as Crown corporations are classified by their economic activity. With these adjustments to Table 5.3, it is possible that over half of the employees in the service sector receive their salary from public accounts.

Resource Base

The economic strength of the North lies in its resource base: forestry, minerals, oil and gas, and hydroelectric power. Each has been developed to satisfy a demand from outside the North. The occurrence of these resources varies across the North, and the pattern of development reflects this variation. The Arctic has a much smaller resource base than the Subarctic, and little resource development has occurred there. Without a broadening of its commercial resource base, economic growth in the Arctic is uncertain. Our examination of each resource looks at three themes. One is the economic importance of the resource industry to the North and to Canada; the second is the regional concentration of particular resource industries; and the third features regional location factors. Regional location factors include the distribution of the natural resource and its accessibility to world markets.

The Forest Resource

The northern coniferous or boreal forest of the Subarctic, occupies around 80% of the forested area of Canada. It forms a continuous 'green' belt from Newfoundland to the Rocky Mountains and northwest to Alaska. Toward its northern limit, the boreal forest consists of open forest with trees growing farther apart and smaller in size. In the southern part of the boreal forest a denser, closed forest is found, and almost all commercial logging takes place in this part of the boreal forest.

The modern forest industry harvests enormous quantities of timber and pulp wood from the boreal forest, helping to make Canada the world's leading exporter of forest products. By the late 1980s, Canada's forests supported an annual harvest of about 175 million cubic metres. Approximately 70% of this harvest produces building products while the remainder is used in the pulp and paper industry. The annual value of forest production and manufacturing was over $20 billion and most of this output was destined for foreign markets, particularly the United States. While the Subarctic accounted for much of the timber and pulp wood, most building products are produced in southern Canada.

Logging for timber began as early as the nineteenth century in the Subarctic forest zones of Quebec and Ontario. These northern logs were dumped into southward flowing rivers and at key points, such as the confluence of the Gatineau and Ottawa rivers, saw mills were located. After World War I, the demand for paper grew and a number of pulp and paper mills were built in eastern Canada. A similar pattern of forest development occurred in western Canada. The first use of western timber was for lumber, and later as the raw material for pulp and paper plants.

World demand for pulp and paper is increasing. Yet, prices can vary over the short term because of falling demand associated with a global recession. A recession normally leads to a decline in demand and a surplus of production. During more prosperous times, the usual 'development' pattern sees new plants built (often with generous public assistance). The next phase of the business cycle is the inevitable economic slowdown, resulting in over production and layoffs at the least efficient mills.

In the 1980s, a new cycle of pulp and paper plant construction began in western Canada, especially in Alberta. With the discovery that aspen produces excellent wood fibre, companies were attracted to the aspen stands of northern Alberta, Saskatchewan, and Manitoba. The Alberta government, seeking to diversify its economy, leased vast northern hardwood forest lands to five pulp and paper companies, including two Japanese firms, Daishowa Canada Company, Ltd and Alberta-Pacific Forest Industries Inc. By 1990, the Daishowa plant at Peace River was nearing completion and the proposed Alberta-Pacific Forest plant at Athabasca, delayed by environmental concerns about toxic wastes discharge into the Athabasca River, was approved by the Alberta government.

Most older, less efficient pulp and paper plants are located in Quebec and Ontario. Since the forestry industry is under great pressure to reduce its pollution of the environment, these mills face massive expenditures and may have to close. Ironically, almost all criticism is levelled at the proposed new mills, which will produce far fewer harmful effluents for aquatic life than do existing mills.

The importance placed on forest production is reflected in the amount of forest land reserved for non-commercial uses. Less than 3% of Canada's forest lands has been set aside for parks and nature reserves. Similarly, though Indians and Métis have lived for centuries in the North, they control little of this land and their participation in the forest industry has been minuscule. In the past, the ownership issue did not pose a major problem because Indians and Métis used the forested lands for hunting and trapping. Now most of these lands are leased to forest companies and their logging activities conflict with hunting and trapping. Only in northern Quebec do Indians control large blocks of timber lands. This land base was obtained with the signing of the James Bay and Northern Quebec Agreement of 1975.

Mining

Canada leads the world in value of mineral exports. Northern mines play a significant role in mineral exports, particularly in uranium, lead, zinc, nickel, copper, gold, and silver. The mineral industry has also been instru-

Figure 5.2
Major Producing Mines, 1991

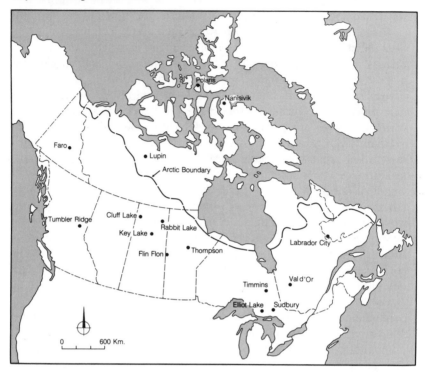

The resource industry relies heavily on mineral production, and important mines and mining towns are found in all areas of the North.

mental in expanding Canada's resource economy into the North. This industry has discovered and developed rich ore deposits in the northern frontier. In doing so, they have had to build company towns for their workers, design their mining operations to fit the northern environment, and ensure a flow of supplies into the community and of mineral product to world markets. All of these activities have strengthened the northern settlement pattern and transportation system.

Mineral production is divided into four sectors: metallics, non-metallics, mineral fuels (oil and gas), and structural materials. Metallic minerals and mineral fuels account for the bulk of the mining activity in the North. Very few structural materials (sand, gravel, clay, and lime) and non-metallic minerals (clay, potash, salt, sulphur, and gypsum) are produced in the North. These minerals tend to be bulky, low-valued products which are very sensitive to transportation costs and therefore must be located near major markets. One exception is asbestos. This product commands a high price and can therefore withstand the high cost of transporting the product to market. Asbestos is mined at Cassiar in northern British

Vignette 5.3   Mineral Development in Northern Quebec and Labrador

Iron ore deposits in northern Quebec and Labrador were developed in response to a growing demand from the United States. Production from the rich Mesabi ore deposit in Wisconsin was running low, and United States iron and steel plants began to look abroad for alternative sources. The previously worthless ore bodies of the Labrador Trough, known for over a half century, suddenly became a valued resource. By 1947, plans were laid for large-scale open pit mining around Knob Lake in northern Quebec. Several mining towns were built, including Gagnon, Schefferville, Labrador City, and Wabush. Schefferville was connected by a 571-kilometre railway to the port of Sept-Iles on the north shore of the St Lawrence River. In 1954, high grade iron ore began to be shipped by barge from Sept-Iles to the iron and steel plants in the Mid-west United States. Most of the capital for this enormous project came from the six large American iron and steel companies. These same companies now had a reliable supply of iron ore for their steel plants in Ohio and Pennsylvania. With a downturn in the demand for iron ore in the United States, production levels achieved in the 1970s fell by half by 1983, and several mines were closed. The days of rapid growth were over and the first serious decline had struck this area of the northern hinterland.

Columbia, and transported by a highway built under the Roads to Resources program to water transport at Stewart.

Metallic mineral production is by far the most important mining operation in the North. It accounts for most of the value of production and employment. Gold is produced at the Lupin and Yellowknife mines; iron ore at Labrador City and Wabush mines in Labrador; lead/zinc at the Faro, Nanisivik, and Polaris mines in the territories; nickel at Thompson, Manitoba; and uranium at Cluff Lake, Key Lake, and Rabbit Lake mines in northern Saskatchewan.

The mining sector is highly dependent on export markets. In 1989, nearly $20 billion of mineral products were exported and over half of this amount came from northern mines. Cyclical downturns in the world economy can force mine closures and lay-offs of miners. During the 1980s, prices for mineral products were low and the mining industry underwent a reduction in its workforce. From 1981 to 1986, the number of people living in 115 mining communities across Canada declined by nearly 3% (Armstrong and Kendall 1990). The greatest decline took place in northern Quebec where the iron mines at Gagnon and Schefferville closed. While neighbouring mines at Labrador City and Wabush still operate, the prosperous economic times associated with the development of vast Labrador Trough iron deposits during the 1970s appears over.

In the two territories, where alternative resource development is slight, the mineral industry is particularly important. In 1989, the value of territories' mineral production was nearly $1.5 billion. This heavy dependence

Table 5.4
Value of Mineral Production in the Yukon and Northwest Territories, 1979 to 1989*
($ millions)

Year	Yukon	% Change	Northwest Territories	% Change
1979	299.2		383.2	
1980	361.3	20.8	367.5	− 4.1
1981	235.6	− 34.8	397.0	8.0
1982	167.8	− 28.8	569.2	43.4
1983	63.0	− 62.5	557.2	− 2.1
1984	70.1	11.0	738.1	32.5
1985	60.1	− 14.3	641.5	− 13.1
1986	176.1	193.0	668.5	4.2
1987	447.2	154.0	810.0	21.2
1988	492.2	10.0	826.5	2.0
1989p	437.2	− 11.2	985.6	19.3

p = preliminary figures

* Excludes value of oil and gas production
SOURCE: Statistics Canada, *Canada's Mineral Production*, Catalogue 26–202.

on the mineral industry means that these economies are vulnerable to sudden declines in price or production (Table 5.4). It also implies that the long-term stability of the economies is uncertain because of the non-renewable nature of this production.

Governments are constantly under pressure to assist the resource industry. The Faro mine in Yukon provides such an example. For nearly fifteen years, the economy of Yukon was heavily dependent on this mine. In mid-1982, the mine ceased production and the Yukon economy went into a tailspin. The value of mineral production dropped from $361.3 million in 1981 to $63 million in 1983 (Table 5.4). By the winter of 1985, only 70 of the original residents remained in Faro. Yukon Government pressured the federal government to get the mine reopened. After a four-year shut-down, Curragh Resources Corporation agreed to reopen the mine, subject to federal and territorial assistance. Financing for the reopening of the mine was heavily guaranteed by the territorial and federal governments. As well, both levels of government provided $8.4 million in incentive grants. The new company recognized that transportation savings could be achieved if larger trucks were used to haul ore to the sea port at Skagway. The federal government agreed to upgrade the highway, while the territorial government committed itself to maintaining the road.

Oil and Gas

During the last thirty years, the petroleum industry has focused much of the search for oil and gas deposits in the Arctic frontier, and discoveries

Table 5.5
Dingwall's Estimates of Canada's Frontier Oil and Gas Potential

Region	Oil (Billion Barrels)	Natural Gas (Trillion cubic feet)
Arctic Islands	1.0	32
Beaufort/Delta	3.5	36
Mackenzie	0.8	4
Labrador	n/a	12
Hudson Bay	n/a	n/a
Offshore Nfld	4.5	9
Offshore Nova Scotia	.3	13
Offshore B.C.	n/a	1

SOURCE: Hladun 1990: 8.

in the Mackenzie Delta and Beaufort Sea have greatly increased Canada's petroleum reserves. These oil and gas fields are not yet viable, particularly because of their high cost of production in a cold environment and the cost of shipping to distant world markets. In the 1970s, there were two proposals to transport natural gas by pipeline from the Arctic Islands to American markets. Both projects failed to materialize because the price of Arctic gas was not competitive in the American market (Vignette 5.4).

Canada's petroleum reserves are found in some 40 sedimentary basins that underlie about half of the land area of Canada and its continental shelves. While all have the potential to contain commercial deposits of hydrocarbons, discoveries of oil and natural gas have been restricted to a few basins, particularly the Western Canadian Sedimentary Basin. This basin, extending from the Beaufort Sea to the 49th Parallel, is the most important petroleum-producing sedimentary basin in Canada. It covers 1.8 million square kilometres, and consists of four smaller basins. Oil and natural gas deposits have been found in its northern basin, northeastern British Columbia, northern Alberta, and the Northwest Territories. These deposits make up approximately one-third of the hydrocarbon production from the Western Canadian Sedimentary Basin. In the future, the North is expected to account for a higher proportion of this basin's output. Canada's greatest potential for future sources of oil and gas is found in the Arctic basins, particularly the Mackenzie Basin which lies beneath the Beaufort Sea and the Sverdrup Basin which is found in the Arctic Archipelago. Two proposed energy developments, the Mackenzie Delta Gas Project and the Amauligak Project, located in the Mackenzie Delta/ Beaufort Sea region, are expected to trigger further oil and gas projects. The output of Mackenzie Delta Gas Project would account for about 15% of the total gas currently being produced in Canada. Three multinational oil companies (Gulf, Imperial, and Shell) propose to invest about $4.7 billion for gas development and $4.5 billion for a pipeline (Hladun 1990: 9).

More than 200 wells have already been drilled in the Mackenzie Delta-Beaufort Sea region. Approximately one-third of these wells were drilled from ships or floating platforms. By the early 1980s, there were nearly 200 billion cubic metres of recoverable natural gas and around 120 million cubic metres of recoverable crude oil (Nassichuk 1987: 279). The potential reserves are thought to exceed the proven reserves by perhaps as much as 20 times (Procter, Taylor, and Wade 1984). In 1984, the discovery of the Amauligak 'elephant' field gave some indication of the potential size of the Beaufort hydrocarbon deposit. This single discovery doubled the proven oil reserves of the Beaufort Sea deposit. Two years after the discovery of the Amauligak field, a token shipment was sent by tanker to Japan. The significance of this shipment was twofold: that oil could be produced commercially from the Beaufort Sea deposit, and that an arctic tanker route could be used to ship its oil to world markets.

In the next decade, oil and natural gas deposits in the Arctic Islands and the Beaufort Sea-Mackenzie Delta area may be exploited if prices stabilize around $30 US a barrel. Since there are significant variations in production costs, Beaufort oil and gas is likely to be developed first. Even at this field offshore production costs are expected to be considerably higher than those onshore. Some indication of this difference is provided by exploration costs. The onshore drilling costs are approximately five times higher than those in Alberta while offshore drilling costs are around 20 times higher (Procter, Taylor, and Wade 1984: 24).

Water Power

The rivers of the Canadian North contain enormous potential for hydroelectric power development. For many years, this potential power was unused because of the high cost of transporting electrical energy to major industrial areas and because alternative energy sources were available at the market. Increased demand for energy and technological advances in electric power transmissions have made northern hydroelectric projects viable. Over the last hundred years, the cost of transmitting electrical energy over long distances has decreased and the technology to build transmission lines has advanced. In 1903, for example, the transmission of electrical power some 140 km from Shawinigan Falls to Montreal was considered an engineering feat. Some eighty years later, Hydro-Quebec's James Bay transmission lines to southern Quebec are almost ten times longer.

Coupled with the ability to transmit energy long distances, the price of electrical energy in southern Canada and the United States increased sharply. Much of this increase occurred after 1972 when the Organization of Petroleum Exporting Countries (OPEC) began to raise the price for their crude oil. By the late 1970s, utilities in energy-deficient New England

Vignette 5.4 The Polar Gas and Arctic Pilot Projects

The Polar Gas Project, formed in 1972, was financed by TransCanada PipeLines, Panarctic Oils Ltd, Ontario Energy Corporation, Petro-Canada, Pacific Lighting Gas Development Company, and Tenneco Oil of Canada. Their first proposal, in 1977, called for the construction of a pipeline along the Hudson Bay coast. Their 1984 proposal called for the construction of a natural gas pipeline from the Drake Point field located in the northern part of Melville Island to service the Ontario, Quebec, and neighbouring United States markets. This 3,763–kilometre pipeline would interconnect with the TransCanada PipeLines system just east of Lake Nipigon in northern Ontario. The proposal allowed for a spur line to connect to the Beaufort natural gas fields. The main pipeline route would begin on Melville Island, cross to Victoria Island, and reach the mainland near Coppermine. It would then travel in a southeasterly direction into northern Saskatchewan east of Lake Athabasca and cross the Subarctic of Manitoba and Ontario.

The construction of this large-diameter pipe would present two major engineering challenges. The first would be to construct a gas pipeline in the zone of continuous permafrost. Because the heat generated by the flowing gas would destabilize the permafrost and therefore threaten the integrity of the pipe, there are two possible solutions. One would be to build an elevated pipeline, and the other would be to chill the natural gas below the freezing point. The second challenge would be to lay pipe on the bottom of the Arctic Ocean. The longest underwater stretch would be the crossing from Melville Island to Victoria Island. The length of this crossing beneath the ice-covered waters of Viscount Melville Sound would be about 150 kilometres. The successful completion of this pipeline would be a major engineering feat.

In 1981, the Arctic Pilot Project proposed an alternative way of transporting natural gas from the Arctic Islands to southern markets. Sponsored by Petro-Canada, Dome, Nova, and Melville Shipping, the Arctic Pilot Project advocated the use of ice-breaking liquefied natural gas tankers (LNG). This plan called for two LNG tankers to collect gas from Melville Island and deliver it to a re-gasification plant either near Quebec City or at Melford Point on the Strait of Canso. This gas would supply the Quebec and Atlantic Canada markets. The ice-reinforced LNG tankers would follow the Parry Sound route by entering through Lancaster Sound, crossing Barrow Strait and Viscount Melville Sound to Melville Island. While this sea route may be ice-free during the short summer period, it would be covered by thick ice during the long winter period and the tankers might well encounter pack ice in Parry Sound and icebergs in Baffin Bay.

were searching for alternative energy sources and Quebec's James Bay hydroelectric project became a major supplier of electrical energy.

Most of the developed hydroelectric power has occurred in the Cordillera and the Canadian Shield. Ideal sites have two characteristics: a large volume of water, and a steep drop. Potential hydroelectric sites still exist in northern British Columbia, Quebec, Manitoba, Newfoundland, and Saskatchewan. Ontario, which has the greatest need for more energy, has only minor hydroelectric sites in its share of the Canadian Shield. This irony of geography is due primarily to the relatively low elevations found in Ontario's Canadian Shield, and its northward flowing rivers.

Historically, small hydroelectric sites in the Shield were developed in conjunction with other industrial developments. These industries included pulp and paper plants and smelters for aluminum, copper, and nickel. These industries require large amounts of energy to process their raw material, and hydroelectric plants offered low-cost energy. Power, therefore, was and is a critical factor in determining the location of these manufacturing plants.

Large hydroelectric projects require large water reservoirs. To smooth out fluctuations in river flows, these reservoirs are sometimes supplemented by storage dams and water diversions. Ensuring a reliable flow of water throughout the year maximizes the installed generating capacity of the hydroelectric facility. Even with an involved system of reservoirs, storage dams, and diversions, a long period of below-average precipitation can reduce river flow and hence electrical power generation. In the 1980s, the Subarctic experienced below-average precipitation and hydroelectric power production at installations in northern Manitoba and Quebec was adversely affected.

The most massive water-power development project in the world is taking place in northern Quebec. The James Bay Project, announced in 1971 by Quebec Premier Robert Bourassa, is a key component in Quebec's economic development. The James Bay Project calls for the harnessing of the rivers flowing eastward into James Bay. These rivers originate in the uplands of the Canadian Shield in the interior of northern Quebec. The first phase, James Bay I, began in the early 1970s and cost $15 billion. It involved massive diversions of water from three rivers (Eastmain, Opinaca, and Caniapiscau rivers) to five dammed reservoirs on La Grande Rivière, increasing the river flow from 1700 to 3300 cubic metres. Today James Bay I consists of three river basins, five water reservoirs, two river diversions, eight dams, three powerhouses, and nearly 200 dikes. La Grande-2 powerhouse (LG2), completed in 1982, is the world's largest underground power facility. Both La Grande-3 powerhouse (LG3) and La Grande-4 powerhouse (LG4) were producing power in 1984. In 1991, another phase was announced and power yet to be produced sold to American utility companies. James Bay II, like its predecessor, involves river diversions, storage dams, powerhouses, and numerous dikes. It is opposed by environmental organizations and Native groups.

The James Bay Project has overshadowed other hydro developments in the North for two reasons. First of all, the massive nature of the project is staggering. Secondly, it is generating more than 10,000 MW annually. Third, the project has reshaped (some say destroyed) the landscape of the three river basins. Fourth, the project has had serious impacts on the Indian and Inuit peoples of northern Quebec and in 1975 the Quebec Government agreed to a land settlement with the Cree and Inuit.

Some ten years earlier, another mega-project was undertaken in northern British Columbia. In 1948, Alcan proposed to develop a complex at Kitimat, which would serve as a port, smelter, and townsite for processing bauxite, principally from Jamaica, into aluminum. Since the smelting of bauxite into aluminum requires enormous amounts of energy, the attraction of this site was the existence of enormous water supplies which could be converted into hydroelectric power. The waters of the Nechako River were dammed and forced to flow in a westerly direction. At Kemano, a tunnel was drilled through a mountain, allowing the water to drop to near sea level. By 1951, construction of a hydroelectrical generating station was begun. In 1978, Alcan initiated studies of a diversion of the Nanika River. This diversion would increase the flow of water into the Kemano generating station by over 80%, allowing enough additional electrical power to be generated to support two more aluminum smelters (Rosenberg, Bodaly, Hecky, and Newbury 1987: 76). Because of pressure from environmental and Native groups, this $600 million hydroelectric power expansion at Kemano may not be completed by 1994 (Howlett 1989: B3).

Resource Towns

Resource development in the Canadian North has led to the establishment of many single-industry towns (Robinson 1962; Stelter and Artibise 1982). Located in remote areas, these towns play an essential role in Canada's resource-based economy by housing the employees working for the resource company. These towns also provide a base for further development in the region.

For the majority of Canadians, resource towns are not attractive places to live (Gill 1986: 23; Siemens 1973). Social life in these communities was described by Lucas (1971: 37) as generating 'feelings of dependency, powerlessness, resignation and fatalism'. The design of resource towns has since improved (Robinson 1962; Lucas 1971; and Gill 1989), and most post-1970 resource towns contain the same amenities as in similar size towns in southern Canada. These changes have greatly improved the social and family environment of single-industry towns. Resource centres with more than 5000 inhabitants can support a range of retail and professional services considered essential by most Canadians. For this reason, larger resource towns better meet the expectations of workers and their families. In spite of these improvements, residents of single-industry towns remain troubled by the presence of the company in their lives. They also consider their stay in the community as 'temporary', have a sense of

Table 5.6
Population Changes in Northern Mining Communities Affected by Mine Closure or Reduced Production, 1981 to 1986

Community	1981	1986	Change	%
Faro	1,652	400	1,252	− 75.8
Labrador City	11,538	8,664	2,874	− 24.9
Leaf Rapids	2,356	1,950	406	− 17.2
Lynn Lake	2,142	1,665	477	− 22.3
Schefferville	1,997	322	1,675	− 83.9
Uranium City	2,507	171	2,336	− 93.2
Wabush	3,155	2,637	518	− 16.4

SOURCE: Statistics Canada, 1987. *Profiles: Census Divisions and Subdivisions.*

isolation from the rest of Canada, and feel uncertain about the future of their jobs.

A resource town's dependence on a single industry can be disastrous if that industry ceases to function. A few have managed to evolve into regional centres with a more diversified economy, but many fail to survive for three main reasons. The key factor in failure is that the town has no other economic function. A second factor is a remote location, making it difficult to attract other industries. Another factor is most Canadians' preference to live in urban centres in more temperate climates than those found in the North.

During the 1980s a number of mines were closed, with severe impact on the local communities (Table 5.6). In 1982, the Crown corporation Eldorado Mining and Refining Ltd announced that it was closing its mining and milling operations near Uranium City. Without any hope of finding an alternative industry, Uranium City experienced economic collapse and depopulation. By 1986, its population had dropped to less than 200, and the town was effectively abandoned. In 1983, Schefferville suffered a similar mine closure. When the Iron Ore Company of Canada ceased its iron mining operations, the population of this single-industry town dropped from nearly 2000 in 1981 to 322 in 1986. In the late 1980s, other mines closed, including the lead-zinc mine at Pine Point, Northwest Territories, and the copper mine near Leaf Rapids, Manitoba. Elliot Lake may suffer a similar fate if the nearby uranium mine is closed.

Not all resource towns suffer the same fate. A few have grown into regional centres with a more diversified economic base. Yellowknife owes its beginnings to the discovery of gold in 1934, but Yellowknife's future was secured when it was selected as the capital of the Northwest Territories in 1967. With a population of approximately 12,000 in 1990, Yellowknife supports a number of service industries supplying the city and surrounding communities with a variety of goods and services. Val-d'Or, Quebec was established in 1935 after a major gold strike was made in what turned out to be the richest gold-bearing ore field in Quebec. Soon, a number of mines were opened. Even though some of these mines have closed, the town has prospered and grown. Val-d'Or, with a population of around 23,000 in 1990, is now an important regional centre with a mining, lumber, and service-based economy.

Single-industry towns designed around a non-renewable resource have a limited life span. Miners and mining companies are hardened to this fact, and this psychological mind-set can lead to a 'migratory' outlook where families avoid making commitments to their community and the North. When mines do close, workers usually receive compensation, which softens their dismissal. As well, miners usually live in company housing and so do not lose real-estate equity. Those people who are financially hurt by the closing of a mining town often own small businesses supplying a variety of services to local residents, and/or own houses which now have lost their market value. The bitterness felt by many residents of a single-industry community when a company closes its mines has been noticed by both companies and governments. For some 25 years, some companies have used an alternative approach: the transporting of workers to and from remote sites.

Air Commuting

An alternative approach to establishing resource towns is air commuting, a new form of the suburban-to-city 'journey to work' model. Air commuting involves much longer travel distances than the traditional daily journey to work, and it requires living at the worksite for several weeks to a month or more. Companies and governments are attracted to this approach because it avoids the costly business of building a new town and the messy business of closing it down when the mining operation is completed.

Air commuting normally involves transporting workers on a weekly or monthly basis from a major city to the mining site which has accommodation and a recreational complex for its shift workers. Air commuting began in the 1970s and continues to be popular with companies operating remote mines. In the case of companies in Saskatoon, the long distance air travel is to three uranium mines located in northern Saskatchewan (Figure 5.3).

Air commuting has serious implications for northern development because most of the jobs are transferred to cities located in southern Canada, leaving few economic benefits in the North. Such economic leakage does not promote northern development, because a southern metropolis captures the bulk of the economic impacts of northern resource development. A partial solution to the problem involves creating a complementary air-commuting system in the North. Run in conjunction with the southern-based air-commuting system, the northern system allows the North to capture a share of jobs and wages. Besides creating more high-paying jobs in the North, the northern air-commuting system

Figure 5.3
Air-Commuting Routes to Uranium Mines in Northern Saskatchewan

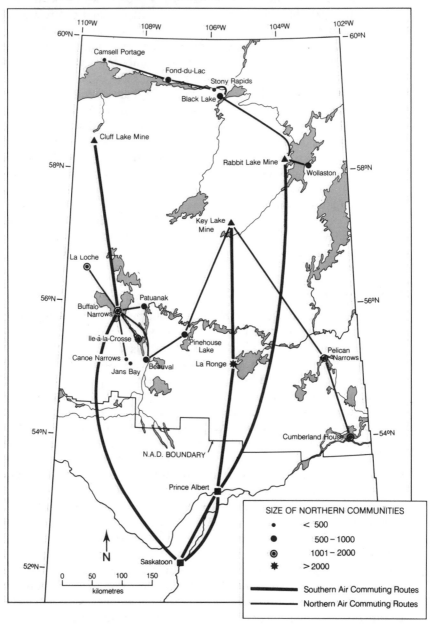

spreads job opportunities over a wide number of small communities and injects more spending power in northern communities. Local communities benefit from increased spending power and a modest level of local multiplier effects, which might lead to additional rounds of economic benefits, i.e., more employment in retail shops and in construction work.

Governments of Saskatchewan and Northwest Territories have encouraged companies using a fly-in labour force to couple their southern-oriented air system with a northern one. In 1975, the NDP government of Saskatchewan stipulated in the surface lease to the Cluff Lake mine operators that half the labour force must be from northern Saskatchewan. The succeeding Conservative government subsequently relaxed that requirement but the uranium mining companies continued to operate northern air-commuting systems. Companies with northern operations realize that employing northerners is an important element of doing business, and Canadian Mining & Energy Corporation (Cameco) announced in November 1990 that by 1995 it plans to increase the proportion of northern workers at its two uranium mines in northern Saskatchewan from around 35% to 50%.

In the Northwest Territories, the Lupin goldmine, located just south of the Arctic Circle near Contwoyto Lake, began its air-commuting system by flying its employees from Edmonton to its arctic mine in the barrenlands. Later, in order to attract northerners to its work force, the company established air charter services between Coppermine and its mine site, adding some 30 residents of the Northwest Territories to the labour force at the Lupin mine (Storey and Shrimpton 1989: 17). The success of its Coppermine fly-in system has caused the company to expand its northern air-commuting system to include Cambridge Bay.

The cost of the northern air-commuting system can be justified in two ways. Firstly, it can be treated as a cost of doing business in the North. If these northern transportation costs increase its total commuting costs, the company's profit will be reduced. On the other hand, if the demand for labour exceeds the capacity of the aircraft used on the southern flight, the addition of small aircraft to shuttle workers from a number of northern communities to the worksite may actually save the company money. This saving would result from reduced transportation costs, assuming that the cost of chartering a small aircraft is less than that of a large one.

Since Asbestos Hill mine in northern Quebec began a fly-in operation in 1972, there have been at least fifteen other such mining operations (Storey and Shrimpton 1989). Only three mines have been operating for 10 years or more. Most mines have closed within five years, suggesting that many companies employing a fly-in system to mines have limited ore reserves. From a company's perspective, an air-commuting system has high annual costs (transportation costs) and low site costs (accommodations for employees), while a single-industry town has a high start-up cost (construction of apartments, houses and a variety of amenities now expected in urban places) and low annual operating costs. The cost trade-off between building a resource town and using an air-commuting system seems to be around 15 to 20 years for a conventional mining operation.

Beyond 15 years, the annual cost of air commuting soon matches the initial costs of constructing a resource town.

Economic Realities, Challenges, and Strategies

In the relatively short time of some forty years, the North has been propelled into a new economic direction, that of a resource hinterland. Most earlier primary industries were small-scale operations found in the southern edge of the Subarctic and in southern Yukon. At that time, much of the Subarctic and all of the Arctic were remote areas far beyond the reach of the global economy. In sharp contrast, today's resource-oriented economy is both a product of the industrial world and dependent on it. Controlled by the world demand for primary products, the resource economy is subject to sharp fluctuations in economic performance. Governments are constantly under pressure to assist the resource industry, particularly in hard times.

One consequence of the new economy is a change in the northern landscape. Resource towns, highways, pipelines, and settlements now form the economic landscape of the North. This landscape has been shaped by external forces, particularly global demand for raw material and energy, and by provincial and federal governments. In this sense, a partnership between the public and private sectors was required to overcome the geography of the North. Governments provided access to resources while private companies, particularly multinational ones, developed the natural wealth. At first, the public sector was reluctant to become involved and only highly valued minerals like gold and the more accessible timber stocks and hydroelectric sites along the edge of the Canadian Shield in Ontario and Quebec were exploited. More remote resources, particularly the bulky, less valuable minerals, had to await public assistance and the development of a northern transportation policy. By the 1950s, the economic reality facing the North was that of a resource hinterland, supplying energy and primary products to the global economy. This economic relation was expressed some 25 years ago by Dunbar (1966: 24): 'Settlement, however, is not based on the ability to survive but on the ability to make a living, and permanent settlement in the North will depend in the long run on demand for the resources of the North.'

Along with the new resource-based economy comes a new set of challenges. One challenge is diversification of the northern economy. Diversification is a slow and difficult process under the best conditions. It is true that regional development in Canada has been initiated by resource export industries. This theme runs deep in Canadian studies and is known as the staple theory. Geographers, however, have noted that all regions are not the same. The North, with its limited resource base and its high cost of exporting products to world markets, is the least well-suited region for this process. One of the major problems preventing economic diversification is the absence of a large northern population. Another is the leakage of secondary effects generated by northern resource industries to southern Canada. By purchasing most of their supplies from southern manufacturers, northern industries transfer secondary effects outside the North. Similarly, air commuting of workers from southern centres to northern worksites transfers spending power to the south. Since the leakage problem occurs in an open economy, the idea of a closed economic system for the North has appeal. Like the argument for 'infant industries', the 'infant region' approach would, in theory, keep the secondary effects in the North and employ northern residents. But with the growing liberalization of trade in Canada as well as the Free Trade Agreement with the U.S., such economic barriers are no longer an acceptable strategy.

An equally serious challenge to the northern resource economy is its heavy dependency on non-renewable resource development. Since non-renewable resources have a fixed life-span, the northern economy is affected by the exhaustion of ore bodies and oil fields. Fluctuations in prices for resources also affect the northern economy. Both these factors lead to an unstable economic environment, characterized by sudden upswings in economic conditions followed by sharp downturns. Exploitation of these deposits can provide an economic boost to the northern economy for a short period of time, but this form of development does not provide a lasting foundation. Such development instills a high degree of instability and uncertainly in the economy. Fluctuating demand for minerals and energy tends to result in a 'boom and bust' type of economy. One consequence is regional economic uncertainty and vulnerability. This uncertainty and vulnerability is caused by world forces far beyond the reach of northerners and other Canadians but the negative impact is felt in the resource communities of the North.

Renewable resources, such as forests and water, do exist in the North. Both are capital-intensive industries. The forest industry employs a great number of northern workers, many of whom are engaged in logging and reforestation activities. Hydroelectric companies, on the other hand, employ few northerners because their electrical generation system is a highly automated operation. While the same degree of automation is not expected in the forest industry, it is increasing. These technological innovations will result in higher productivity per worker and increase the profitability of the operations. They will also lead to a reduction in the size of the labour force.

Finally, the issue of large-scale resource projects in the North is a controversial one. Some have heralded such massive development as the triggering device for a strong northern economy, while others see it as the exploitation of northern resources by outside interests. Mega-projects are expensive and require large capital investments and secure markets. Since large international companies have the necessary resources to undertake large-scale projects, these firms dominate the northern economy. Given the scope of their projects, these corporations are often able to obtain special concessions from the federal or provincial governments. Such concessions may include improving the transportation links to the resource, tax concessions, or loan guarantees. In this way, large companies are able to reduce the risk of the project and enhance its profitability. Public funding of foreign resource companies does raise the question, does the return to Canada and the North justify the subsidies? For Canada, it seems the answer is yes, but for the North a positive answer is more difficult. And yet, what is the alternative?

Control of the resource economy by large firms raises other questions about their impact on the northern economy. While their investments and northern job creations are positive factors, what commitment do they have to the long-term development of the North? With their headquarters in Calgary, New York, Quebec City, Saskatoon, Tokyo, and Toronto, executives are far removed from northern issues and problems. The complex subject of large-scale development is discussed in the next chapter where such development is presented with a case study of the Norman Wells Project.

Selected Readings

Dacks, Gurston, 1981. *A Choice of Futures: Politics in the Canadian North*. Toronto: Methuen.

Gill, Alison M., 1986. 'New Resource Communities: The Challenge of Meeting the Needs of Canada's Modern Frontierpersons,' *Environments* 18(3):21–34.

Rosenberg, D.M., R.A. Bodaly, R.E. Hecky, and R.W. Newbury, 1987. 'The Environmental Assessment of Hydroelectric Impoundments and Diversions in Canada', in *Canadian Aquatic Resources*, edited by M.C. Healey and R.R. Wallace. Canadian Bulletin of Fisheries and Aquatic Sciences 215, Department of Fisheries and Oceans. Ottawa: Minister of Supply and Services.

Storey, Keith and Mark Shrimpton, 1989. *Impacts on Labour of Long-Distance Commuting Employment in the Canadian Mining Industry*, ISER Report No. 3. St John's: Memorial University of Newfoundland.

Waldram, James B., 1988. *As Long as the Rivers Run: Hydroelectric Development and Native Communities in Western Canada*. Winnipeg: University of Manitoba Press.

Wallace, Iain, 1987. 'The Canadian Shield: the Development of a Resource Frontier', in *Heartland and Hinterland: A Geography of Canada*, edited by L.D. McCann. 2nd ed. Scarborough: Prentice-Hall, Chapter 11.

Chapter 6

MEGA-PROJECTS IN NORTHERN DEVELOPMENT

The Canadian North has entered a new phase of resource development: large-scale resource projects–mega-projects–financed and managed by multinational and crown corporations, and designed to meet global needs for primary products. This phase, in which multinational or transnational corporations conduct business across national boundaries, represents the increasing internationalization of production, manufacturing, and trade. Two recent projects, the James Bay Hydroelectric Project and the Northwest Coal Project, demonstrate both the scope of such mega resource undertakings and their export orientation dependency. A dozen such projects–some proposed, others completed–are listed in Figure 6.1.

Definition and Characteristics of Mega-Projects

Mega resource projects are industrial undertakings which, because of their enormous size, dominate the local and regional economy during the construction phase. Construction costs usually exceed $1 billion. Once in operation, the output from these plants greatly increases the value of production for the region and the country as a whole. Mega-projects have a number of common features. For the North, their single most important characteristic is the strengthening of local and regional economies. The most visible economic impact occurs in the community located near the project; in others, an existing place is expanded. In many cases, a new town

Figure 6.1
Mega-projects: Proposed and Completed

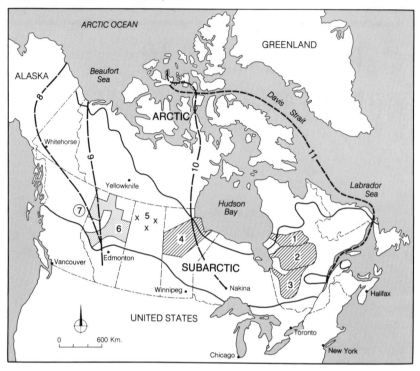

Over the years, a number of mega-projects have been proposed but only a few have been built. The map shows a representative selection.

Proposed

1 Nottaway-Broadback-Rupert Project Phase 4, James Bay Project
2 Great Whale River Project, Phase 3, James Bay Project
8 Alaska Gas Pipeline (from Prudhoe Bay to the United States)
9 Mackenzie oil and gas pipeline projects (from the Mackenzie Delta-Beaufort Sea oil and gas deposits to North American markets)
10 Polar Gas Pipeline Project (from Melville Island gas fields to markets in North America)
11 Arctic Pilot Project (from Melville Island gas fields by tanker to markets in North America)

Completed

3 La Grande River Project, Phases 1 and 2, James Bay
4 Churchill-Nelson River Project
5 Uranium mines in northern Saskatchewan
6 Timber lease areas for proposed pulp plants in northern Alberta
7 Northeast BC Coal Project

is built to house the workers at the project. The size of the investment, the length of construction period (often two years or more), and the proportion of investment allocated to building a transportation system (over half of the total investment for energy projects) mean that most firms undertaking a mega-project are multinational or Crown corporations.

Such corporations have the capital, technology, and managerial skills necessary to undertake large projects of this magnitude. They are often vertically integrated, which means that their resource production has a ready market within their corporate structure or through long-term contracts.

Large-scale resource projects are widespread in the Subarctic and have yet to make an appearance in the Arctic, although several have been proposed. Polar Gas Project and the Arctic Pilot Project both sought to transport natural gas from Melville Island to southern markets, while the Beaufort Sea Oil and Gas Project calls for Mackenzie Valley pipelines. The sheer size of such undertakings alters the regional economy, locking it into an export-oriented hinterland. These mega-projects call for huge investments. Remote energy and base-metal projects require expensive transportation systems. Pipelines or tankers are needed to ship oil and gas to market, high voltage transmission lines must convey hydroelectric power, and railways or tankers must transport base metals. The economic advantage of large-scale projects is that the size of the operation allows economies of scale to be realized in both production and shipping. In this way, the high costs of operating in a cold environment and of shipping the product long distances to market can be offset.

Historic Origins of Large-Scale Resource Projects

The Klondike Gold Rush of 1897–98 was the first large-scale mining effort in the North, but unlike later large-scale resource projects, the Klondike Gold Rush was not corporate-driven, because placer gold mining could be undertaken by individuals. Placer mining involved the recovery of auriferous deposits found in stream beds by simple and inexpensive technology–panning or surface sluicing. This technology enabled seasoned prospectors and greenhorn amateurs to recover the most accessible nuggets and fine sand-like gold particles from tributaries of the Klondike River, particularly Bonanza Creek. Once this source of gold was exhausted, coarse gold buried well below the surface was sought. Since the mining of this gold required much more capital-intensive power dredging and hydraulic operations, along with separate, highly organized water-supply systems and electric power supplies, large companies gradually took over the gold mining industry in Yukon. The Yukon Gold Corporation was the largest of these gold mining operations. In order to supply its mining sites on the Bonanza and Eldorado creeks with water and electrical power, it built a hydroelectric dam and a hundred-kilometre-long water distribution system (Rea 1968: 99–100). This more complex mining was limited to the summer months when the frozen ground could be thawed and the gravel sorted. During the long summer days, the dredges operated around the clock, although the existence of permafrost made hydraulic operations more difficult and time-consuming.

Northern mining typically required much capital and was suited to large companies. The base metal mine at Flin Flon in northern Manitoba came into production in 1929. A large copper-zinc ore body had been discovered in 1915, but the Flin Flon mine took nearly 15 years to develop. The first task was to secure enough capital. The second task was to define the ore body and to assess the ore grade. The assessment revealed that the copper-zinc sulphide ore would require the development of a new flotation and separation process, and so more capital had to be raised to pay for the development of new technology and the construction of a pilot plant at the site. Once this problem had been solved, the mine, concentrator, smelter, hydroelectric dams, and power station, a townsite, and a railway branch from The Pas were built. Except for the CNR railway line, the construction and infrastructure of the mine were paid for by the Hudson Bay Mining and Smelting Company. The company also sold power to a Sherritt-Gordon mine which came into operation in 1931 some 60 kilometres north of Flin Flon, and smelted copper concentrate from the Sherritt-Gordon mine.

During the Second World War, the inland oil field at Norman Wells was expanded and a pipeline built from the oil production site to Whitehorse, where a refinery was built. This mega-project was financed by the U.S. Army, because Washington wanted an alternative supply of oil which could not be impeded by an enemy submarine or aircraft attack. This wartime mega-project was the Canol project (Canol was an acronym for Canadian Oil). Work began in 1942 and was completed in early 1944, by which time the threat of an attack on Alaska by the Japanese had diminished considerably. Without military demand, Canol oil could not compete with lower-priced California oil. The refinery was closed, the new wells at Norman Wells capped, and the pipeline abandoned. By 1947, the pipeline, pumping equipment, and support vehicles were sold as surplus war assets. Imperial Oil purchased the Whitehorse refinery and moved it to Edmonton where it processed oil from the Leduc field.

The era of mega-projects really began after World War II as American industry seeking foreign raw materials and energy triggered a series of large-scale resource projects. The most prominent mega-projects took place in northern Quebec and adjacent areas of Labrador where iron mines, mining towns, a railway, and port facilities at Sept-Iles were built to supply ore to United States steel plants. Starting in the 1970s, foreign investment from other countries but particularly from France, Germany, and Japan was attracted to northern Canada. These countries, like the United States, were seeking a reliable source of raw materials and energy. Oil exploration in Canada is dominated by American companies while American, French, and German firms have developed the uranium mines in northern Saskatchewan. Japanese companies have recently obtained long-term leases for virtually all commercial timber in northern Alberta. Major hydroelectric power developments, on the other hand, are undertaken exclusively by provincial Crown corporations. They, like transnational corporations, are attracted to resources in the North because production and transportation costs are competitive with alternative sources. Hydroelectric Crown corporations may also express a provincial strategy for economic development. Provincial hydroelectric Crown corporations in British Columbia, Manitoba, and Quebec have built northern hydroelectric projects to produce low-cost energy and to sell surplus energy to the United States.

The Role of Multinational Corporations

The cost of developing northern resources is often beyond the financial resources of an individual entrepreneur and small firms. Multinational corporations most often possess the vast capital resources, up-to-date technology, world-scale marketing structure, and aggressive (efficient) management needed to successfully develop large-scale resource projects in the North. With the high costs of exploration, development, and marketing of energy and mineral deposits, these corporations have the best chance of avoiding failure and making a profit from northern resources. But there is always a high risk factor in northern development. The collapse of Dome Petroleum, the rejection of the Mackenzie Valley Pipeline Project, and the shut-down of the iron mines owned by the Iron Ore Company of Canada, provide recent examples of the high risk associated with northern ventures.

The global business cycle can affect the success or failure of such projects. The Northeast Coal Project in British Columbia provides such an example. In the late 1970s, plans were launched to develop the vast coal reserves in northeastern British Columbia at a time when the world economy was expanding and the demand for coal was increasing. Three coalmining operations comprise the Northeast Coal Project: Quintette Coal Ltd (Tumbler Ridge), Teck Corporation (Bullmoose), and Gregg River Coal Ltd (Gregg River). The total investment was about $4.5 billion, with the federal and British Columbia governments providing a total of $1.5 billion to build two railways and port facilities from which to ship coal to Japan. Some fifty banks loaned a total of $1 billion to Quintette Coal Ltd–the largest of the three operations–to develop a mine and the town of Tumbler Ridge. The Japanese steel mills were willing to purchase the entire production of Quintette's operations (nearly 5 million tonnes annually) at prices above the world market in order to obtain a secure supply of coal. World prices, then under $100 a tonne, were expected to rise above this figure. Both Quintette and a consortium of Japanese steel producers were confident that time was on their side. A 14-year contract set the price at $75 a tonne in 1980 but with annual escalations possible to a ceiling price of $104 a tonne. The contract also called for three price reviews, one in 1987, another in 1991, and the last in 1995. Before the mine opened in 1984, the world economy was already sputtering and global coal prices were declining. However, when the Quintette mine opened, the contract price was over $90 a tonne.

In 1987, the contract price for Quintette coal was to be reviewed. At that time, it was nearly $100 a tonne while the world price was just under $60 a tonne. The Japanese consortium wanted the price decreased to world levels while Quintette demanded the fully escalated contract price of $104. The dispute was taken to arbitration and the arbitrated prices were set at about $96 for the period from 1987 to 1990 and at $82.40 for three months thereafter. Assuming that the contract price remains around $82 a tonne after the next price review, the Japanese steel industry will be paying about $22 a tonne below what Quintette claims it costs to produce the coal, deliver it to port at Prince Rupert, and service its $700 million debt. Quintette Coal Ltd may be forced into bankruptcy. In 1990, the major shareholder in this project, Denison Mines Ltd of Toronto, had written off 51% of its investment. In the next upturn, coal prices may again exceed $100 a tonne but by then Quintette Coal Ltd may no longer exist and the physical assets will be owned by the banks. The problems confronting this coal project have promoted a number of cynical interpretations of mega-projects.

Benefits and Costs of Mega-Projects

Large-scale resource projects in Canada's North usually involve foreign investment and ownership. Private companies are attracted to Canada's North for non-economic reasons. These reasons include a stable political environment for private business which ensures that their investment and source of supply are safe, and friendly governments which may offer substantial financial concessions to private developers. But what are the benefits and costs to Canada and its North?

Large foreign firms dominate the resource industry and therefore set the agenda for most resource development in the North. The principal advantage of this approach for Canada and the North is that it speeds up the process of resource development, creating jobs, while three factors (profits, experience, and knowledge) increase the company's capacity to undertake future resource projects in the North and other parts of the world. The chief disadvantages are that profits flow out of the North and Canada, and that the managerial experience and technical knowledge gained from the development is retained by the foreign company. Another disadvantage is that public funds are often used to encourage such developments, and transnational companies are able to reduce their risk and enhance their prospects of a profitable venture at the Canadian taxpayers' expense. A recent example is provided by the Japanese-based Daishowa Paper Manufacturing Co. Ltd. In 1988, it received a commitment for $65 million from the Alberta government and $9.5 million from the federal government (Fisher 1988: B6). The total cost of the hardwood pulp mill is estimated at $500 million, of which public funds represent 15% of the total investment.

Foreign firms are actively involved in the major resource sectors found in the North. In the forest industry, American and now Japanese companies play a strong role. Weyerhaeuser Canada, an American-owned firm, controls most of the commercial forest in northern Saskatchewan through its timber licence. From April 1990 to March 1991, Weyerhaeuser required 2 million cubic metres of wood. Just over 60% of this wood came from harvesting operations on selected areas within its Forest Management Licence Area in northern Saskatchewan and the rest was purchased from private landowners, sawmills, and contractors within the province. In the mining industry, examples of such large-scale development abound. American investment financed most of the iron ore mines in northern Quebec and Labrador, French and German funds helped build the uranium mines in northern Saskatchewan, and Japanese steel firms invested capital in the coal industry of northeastern British Columbia. In most cases, concentrated ore is shipped to processing factories in the home country. A major project under consideration by the federal Environmental and Review Office, the Kiggavik uranium mine near Baker Lake in the Northwest Territories, illustrates this point. The German mining company making the proposal provides an example of the vertical integration of foreign firms. In this case, the uranium oxide would be exported to German power corporations to generate electricity for domestic and industrial consumers. Local opposition to this project has been very strong because Inuit residents are fearful that the mine will adversely affect the wildlife. For example, the site of the proposed Kiggavik uranium mine lies near the breeding grounds of the Beverly and Kamiuriak barren ground caribou.

Mega-Projects: The Engine of the Northern Economy?

According to a 1981 federal Task Force, resource development holds the key for economic growth and northern development. This Task Force, headed by Robert Blair of Nova Corporation and Shirley Carr of the Canadian Labour Congress, estimated that the value of mega-projects expected to be built in the 1980s and 1990s would inject nearly $500 billion into the economy. Close to 80% of the proposed sites of these mega-projects are in the Canadian North (Blair and Carr 1981). The list of projects identified by the Task Force include the now completed Norman Wells Project and the yet-to-be-started oil and gas development in the Beaufort Sea. The 1981 estimated cost of the Beaufort developments by Dome Petroleum was $48 billion. Proposed pipelines in the Arctic were equally expensive. In 1981, Foothills Pipe Lines Ltd, a subsidiary of Nova Corporation, estimated the cost of the Dempster Gas Lateral at $2.5

billion. Mega-hydroelectric projects were also expensive. In 1981, the construction of the Conawapa dam on the lower Nelson River was estimated at just under $3 billion. Ten years later, the cost of these projects would more than double and they would face stiffer environmental reviews. For example, in 1990, the Manitoba Government announced that the construction of the Conawapa Dam at a cost of $5.5 billion would begin in 1991.

Mega-projects not foreseen by the 1981 Task Force are pulp and paper plants planned or under construction across the western Subarctic. These plants, for the most part, hold forest leases on large stands of aspen poplar which is now highly desired for pulp and strandboard. In the boreal forest of Alberta alone, there are seven new pulp mill projects valued at over $1 billion. These pulp plants form the basis of Alberta's diversification strategy for the 1990s. The proposed plants are expected to employ some 12,000 persons, either directly or indirectly (Stirling 1989: 15). Similar but slightly smaller pulp projects are planned for Saskatchewan and Manitoba.

Major construction projects are expected to trigger a burst of regional growth while on-stream projects will have a much longer term but less dramatic impact. At the same time, mega-projects will generate wealth which, if captured by taxes, can be redirected to the financing of social and cultural programs or, if spent by workers and shareholders, can cause a ripple effect through the economy, thus increasing the opportunities for other businesses and workers.

Projects with a ripple effect usually involve the construction of world-scale facilities, such as the proposed James Bay II hydroelectric project, and the Beaufort Sea gas development and pipeline project. The proposed Great Whale hydroelectric project in northern Quebec, set for completion in 1998, is estimated to cost $6 billion (McKenna 1990: B7). Oil-sand projects fall into the same high cost category. Just before the federal election in November 1988, the federal energy minister signed an agreement with a consortium of oil companies to provide financial assistance to the construction and operation of extraction and upgrader plants some 80 km north of Fort McMurray. The project is known as OSLO (Other Six Leases Operations). The consortium consists of Esso Resources, Canadian Occidental Petroleum Ltd, Gulf Canada Resources Ltd, Petro-Canada Inc., Pan-Canadian Petroleum Ltd, and Alberta Oil Sands Equity. Its construction cost is estimated at $4.2 billion. With a 25% share in the project, Esso Resources is the project operator. Under the 1988 agreement, OSLO would have received $850 million in cash from the federal and provincial governments plus nearly $160 million in loan guarantees. Most importantly the companies would have received federal assistance if the price of oil fell below $25 U.S. (Brown 1989: 8). Given that the 1988 price of oil was less than $20 a barrel, this agreement took much of the risk out of the project.

When in February 1990 the federal government withdrew its support, one reaction of the consortium was to relocate the upgrader plant to a more economically viable location, not near Fort McMurray but adjacent to Edmonton. The new plan is to extract the oil-sands and ship them by pipeline to Redwater (a small town near Edmonton) where the upgrader will be located. Constructing the upgrader near Edmonton is expected to save money in two ways. First, using workers from Edmonton who will commute to the construction site at their own expense will reduce construction costs (the alternative is to build a work camp at the mining site). Second, upgrading may be able to operate year-round, obtaining heavy oil from various sources; extraction plants must close for maintenance, several times a year. If the upgrader is located near Edmonton, then much of the economic spin-offs from heavy oil development in the Fort McMurray area will flow out of northern Alberta to the Edmonton area.

Shortcomings in Mega-Projects

There is no doubt that mega-projects have helped develop the resources of northern Canada. But both companies and governments have painted the impact of these project in far too positive a light. In reality, compared to the rest of Canada, the Canadian North receives relatively few benefits from resource projects. Furthermore, there is little evidence to suggest that the northern resource economy will evolve into a more diversified and stable economy.

The factors limiting the economic impacts of large-scale resource projects in the North are related to the market economy and the economic structure of Canada; i.e., while vast sums are spent and many are employed during the construction phase of a northern resource project, most economic benefits flow to southern Canada and other countries. Much of the labour and material needed for the construction and operation of these resource projects comes from southern Canada. As a result, not only does southern Canada benefit directly but it also receives most of the secondary or spin-off effects of northern resource projects (Ironside and Mellor 1978).

These spin-offs are called the multiplier effects (see Vignette 6.1). Ironside and Mellor (1978) found that the indirect and induced employment multiplier from a new forest-product plant at Slave Lake in the early 1970s was low–only 1.2. Assuming a similar multiplier effect on the proposed $1.3 billion Alberta-Pacific pulp plant near the town of Athabasca in northern Alberta, a total of 1,320 jobs is anticipated. Most (83%) are directly related to the mill's operation and its wood supply activities, leaving only 220 'spin-off' jobs (Ironside and Fieguth 1990: 8).

A new form of job leakage, air commuting, may decrease the multiplier effect even further. There is a growing trend by companies to employ an air-commuting system to supply labour to mega-project construction sites.

Vignette 6.1   Regional Multiplier

In neo-classical economic theory, the most common form of economic impact analysis is based on the Keynesian concept of the multiplier. The multiplier is a measure of the economic impact of a new development, such as a factory or mine, on the local or regional economy. There are three types of impacts: the direct impact of the wages, salaries, and profits of the new development; the indirect impact from payments to regional industries supplying goods and services to the new firm; and the induced impact, which is the increase in payments to retail stores and their regional suppliers brought about by the spending of the new income.

The regional multiplier is expressed mathematically as $1/(1-s)$ where s is the marginal propensity to consume goods and services within the region. Goods and services purchased outside the region represent economic leakage. For example, let us assume that the induced impact is determined by a regional multiplier of 1.5. This multiplier indicates that $0.33 of every dollar spent on supplies and wages by the owners of the new enterprise occurs within the region. It is calculated from the expression $(1/1-s)$ where $1/(1-0.33) = 1.5$. Arriving at the total annual income impact involves applying the multiplier to the total expenditures from direct and indirect impacts within the region. If this annual amount was $3 million, then the total impact is $4.5 million ($3 million + 1.5 million).

In the case of the Alberta-Pacific project, the multiplier was used to calculate the number of anticipated indirect and induced jobs (jobs not directly associated with the project). In this example, the multiplier was assumed to be low–1.2–and the number of jobs in the northern region was calculated as follows:

600 direct jobs x 1.2 = 720 total jobs (120 indirect or induced).

If the site is remote, then the company is likely to avoid the expense of building a new town by opting for an air-commuting system. Unless northern centres are also served by this air-commuting system, one consequence is that the labour force resides in southern Canada, and spend their wages there.

Another form of leakage occurs when the construction contract is awarded. The construction firm winning the contract to build the project is usually a large company with previous experience and such companies are found only in the south. Profits gained from the construction work then go to the southern headquarters of these firms.

Another disadvantage facing the North is that large companies have little commitment to the region. While the attitudes of companies may be changing, multinational companies and Crown corporations are content to consider the investment of vast sums in exploration and development of northern resources as their sole contribution. All other matters, they would argue, are the responsibility of local residents and their governments. The main reason for this attitude is the profit goal of corporations; a secondary one may be the location of corporate headquarters.

Profit-making dictates that the economic well-being of the company and its shareholders comes first. The selection of a resource project and the closing of it is based on economic considerations. Resource development by large corporations is directed to global markets, and external market forces determine the existence and life span of the project. For example, a mine may shut down, not because its ore reserves are exhausted, but because the firm has another, more profitable deposit coming into production elsewhere in the world.

The second factor is more problematic. It is based on the argument that the location of corporate headquarters outside the North and often outside Canada places northern resource decisions in the hands of 'outsiders', possibly desensitizing senior decision-makers to local concerns and needs. For example, executives' individual contribution to 'their' community and that of the corporation (which is shaped by the same executives) is rarely directed to the North. On the other hand, executives of companies working in the North have become more aware of the special needs of northern people and a few have authorized the spending of company funds to address these needs. Uranium mining companies based in Saskatoon, for instance, have instituted scholarship programs for northern students. These company programs represent an important shift in attitude–from one that saw the social problems of the North as solely the responsibility of northerners and the federal and territorial/provincial governments, to one that recognizes a role for the private sector.

The Norman Wells Project

Of all mega-projects, the Norman Wells Project was the only one which was monitored during its construction, and it therefore provides details of impact not available elsewhere.

The Norman Wells Project was a major oil and pipeline construction undertaken in the Northwest Territories and completed in 1985. The first energy mega-project to take place in the Northwest Territories after the Berger Inquiry, it cost close to a billion dollars. The project is seen by industry and government as a model for future energy efforts. From an industry perspective, Norman Wells was built without serious environmental or social impacts. As well, it was built some 30% under the original cost estimate of $1.4 billion, and was completed slightly ahead of schedule.

The Norman Wells oil field, discovered in 1920, has proven oil reserves of around 100 million cubic metres of light grade oil (Esso 1980). A small refinery was erected at the site and oil was produced to meet the needs of the residents of the Mackenzie Valley. In 1925, operations ceased because the demand for oil was too small. In 1932, production recommenced because the mining operation at Port Radium on Great Bear Lake demanded fuel oil. In 1936, the gold mine at Yellowknife also required fuel oil. Given the slow growth of a market for oil in the Mackenzie Basin, production increased very slowly until 1944. At that time, the Canol Project was completed and for a short period production increased sharply. With the end of the Second World War, the military market for Norman Wells oil ended and production was again limited to supplying customers in the Mackenzie Valley. From 1946 on, the local market grew through the settling of Native people in houses heated by fuel oil, the expansion of every community's infrastructure (again all heated by fuel oil), and increased mining activities. By the late 1970s, oil prices had risen sufficiently to cause Esso Resource Canada to plan increased production at Norman Wells for shipment to the southern market. This plan took the form of a proposal called the Norman Wells Oilfield Expansion and Pipeline Project (the Norman Wells Project). Esso's proposal called for annual production to increase from just over 180,000 cubic metres/year to nearly 1,500,000 cubic metres (Table 6.1). A water injection technique would increase oil recovery from 17% to 42%. Facilities required for increasing production would include 200 new oil and water injection wells, six artificial islands to serve as drilling platforms, an oil gathering system, and a central processing plant to condition oil for pipeline transmission. The relatively small-diameter (324 mm) Norman Wells line, buried along its entire length, would have a capacity of around 5,000 cubic metres/day. It would transport oil at near ground temperature, thereby reducing the potential for frost heave. Three pumping stations were to be located near Norman Wells, Wrigley, and Fort Simpson.

In 1982, Esso Resources Ltd Canada began the oil expansion program which involved the building of islands in the Mackenzie River and a central facilitating plant. Esso assigned Interprovincial Pipe Line Ltd (IPL) the task of constructing a buried pipeline. The pipe was laid during the winter to minimize damage to the environment. Both Esso and IPL subcontracted the work to other firms, some of which were northern companies.

In 1985, the impact of the new oil production was evident. In 1984, oil production was 175,000 cubic metres ($20 million in value) while a year later it nearly reached 1.2 million cubic metres (almost $200 milion in value). The latest published figure for 1990 indicated a production of 1.9 million cubic metres, worth $250 million (Table 6.1)

The economic benefits stemming from the construction of this energy project occurred at four levels. These are the local area, principally Norman Wells and, to a much less degree, Fort Norman, Wrigley, Fort Simpson, Fort Franklin; many other communities in the Mackenzie Valley and around Great Slave Lake; the Province of Alberta but particularly the City of Edmonton; and the nation as a whole. Local benefits took the form of employment and business contracts in the four impact communities. At Norman Wells, Esso acted as the general contractor and tendered a number of small contracts to non-union sub-contractors and businesses. Some jobs and contracts were awarded to people in other communities

Figure 6.2
Norman Wells Pipeline Route

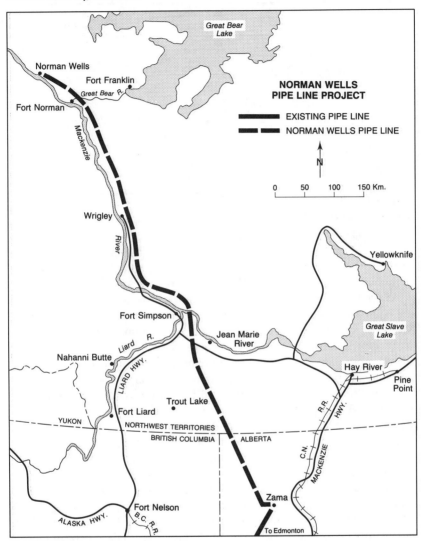

SOURCE: Bone 1988 5: 63.

such as Fort Franklin and Hay River. The hamlet of Norman Wells also benefited by selling rock from its rock quarry to build docking facilities and the artificial islands in the Mackenzie River. Spin-off effects also occurred in Norman Wells with construction of some 30 houses, several buildings, and a school. Most house construction was for Esso employees relocated from company property now required for the new project and for the new employees needed to operate the fieldgate and oil pumping

Table 6.1
Production and Value of Norman Wells Petroleum, 1981 to 1990

Year	Oil Production (000 m³)	$ Value (millions)
1981	172	13
1982	173	15
1983	169	19
1984	175	20
1985	1,148	195
1986	1,478	103
1987	1,570	145
1988	1,833	124
1989	1,885	178
1990*	1,918	250

*Preliminary figures

SOURCE: Statistics Canada, 1991. *Canada's Mineral Production 1991.* Catalogue 26–202.

station. At the completion of the project, Esso and IPL increased their permanent workforce by some 120 additional workers.

At the provincial level, Alberta and other provinces provided most of the equipment and materials required for the construction of the Norman Wells Project. The fieldgate was prebuilt in Edmonton and assembled in Norman Wells. The refinery at Norman Wells processes around 6% of the oil production while refineries in Alberta, Ontario, British Columbia, and Quebec processed the remaining oil. The major contractors for the Norman Wells Project were southern firms. Partec-Lavalin and Northern-Loram were the principal contractors for Esso at Norman Wells while Pe-Ben Pipelines (1979) Ltd and Majestic Contractors undertook the pipe-laying work for IPL. Northern-Loram (a joint venture firm of Northern Construction Company of Vancouver and Loram International of Calgary) obtained a $100 million contract to conduct drilling, blasting, and hauling of rock from the Norman Wells quarry to build a road, dock, and six artificial islands (Bone and Mahnic 1984: 57).

Tax benefits of this energy project accrued to the federal government by means of corporate and personal income taxes. These two sources of tax revenue plus a one-third share in the project's profits were estimated to generate around $172 million additional tax revenue per year (FEARO, 1981: 22). The drop in world oil prices has reduced these projected revenues somewhat but nevertheless the federal government has increased its tax base. Initially the Government of the Northwest Territories received little additional tax revenue because mineral resources belonged to the federal government. Preliminary estimates of GNWT tax revenue amounted to 3% while the federal government would receive 97% (FEARO 1981: 22–3). Under the Northern Accord agreement, energy tax revenues will accrue to the Government of the Northwest Territories.

The nation has reduced its reliance on imported oil. Another benefit of the project to the nation is more efficient use of a non-renewable resource. Prior to the Norman Wells Project, the rate of recovery of oil was about 17%. With the implementation of a waterflood system, the recoverable reserves are expected to reach 42% (Esso 1980). At current production rates the recoverable reserves would be exhausted in 26 years. However, production can continue beyond that time if additional reserves are discovered or if new technology and/or higher oil prices allows the economic recovery of more oil.

The Norman Wells Socio-Economic Impact Monitoring Program

After the federal government approved the Norman Wells Project in 1981, the Department of Indian Affairs and Northern Development was responsible for ensuring that the anticipated economic and social impacts remained manageable. According to the report of the Federal Environmental Assessment panel (Duffy 1981: 4), 'the project impacts on society can be made to be within acceptable limits and the panel recommendations are aimed at minimizing social disruptions.' The Department perceived the need for a monitoring program to collect and evaluate changes in selected statistical data over time. With the permission of settlement and band councils, these data would be collected annually from residents and businesses found in the four communities located near the proposed pipeline route, namely Norman Wells, Fort Norman, Wrigley, and Fort Simpson. A summary of community responses to these surveys was presented to these councils and bands each year for their information and use.

Early in 1982, the Department contracted with the University of Saskatchewan to design and conduct a socio-economic impact monitoring program. This monitoring program was the first in Canada to measure socio-economic changes at the community level as a project was taking place. Its design was simple. The responses of local residents provided annual data on a wide variety of socio-economic and demographic topics. By conducting the same survey each year, changes at the community level could be identified and compared to pre-construction conditions. The first survey took place in 1982 and, except for Norman Wells, construction work had not yet commenced. The data were collected annually by community surveys. During each survey, residents were asked to answer a series of questions about the Norman Wells Project and its impact on them. A similar survey was conducted of businesses and public agencies every other year. The results of this monitoring program were published by the Department of Indian Affairs and Northern Development.

The Major Findings

During the public hearings into the Norman Wells Project, two main concerns had emerged. These concerns focused on possible negative impact on Native people and the size of economic benefits accruing to the North. Both concerns were linked to the expected large influx of southern workers. During the construction of the much larger Trans-Alaskan Pipeline an enormous in-migration of workers had had significant negative economic and social impacts. Strong (1977) estimated that around 60% of the work force for this pipeline moved to Alaska, creating pressure on housing, public services, and community resources. These 'boom' conditions temporarily changed the nature of life in communities along the Trans-Alaskan Pipeline, leaving in its wake a number of negative impacts on Alaskans, particularly Native Alaskans. The Norman Wells Socio-Economic Impact Study was designed to assess any similar such impacts and to address those issues raised in the public hearings held by the Federal Environmental and Assessment Office and by the National Energy Board.

Size of the Labour Force

During the construction phase, the size of the workforce at Norman Wells did increase. In 1981/82, the active labour force totalled 685 persons; in 1982/83, it rose to 1,122; in 1983/84, it reached a peak of 1,498; and in 1984/85, it dropped to 729. The changes in the size of the labour force at Norman Wells reflected the level of the construction work which was greatest in the summer of 1984. While some workers did live in Norman Wells, all major contractors flew workers into Norman Wells on a temporary basis. After two weeks to a month, the workers were flown back to their home community. This approach meant that most workers' families remained in other northern communities or in southern cities. For this reason, Norman Wells did not suffer from large numbers of in-migrations during the construction period.

Demographic Impacts

The population size of Fort Norman, Wrigley, and Fort Simpson did not change significantly during the construction of the Norman Wells Project. In 1981, these three centres had a combined population of 1,403 and, by 1985, it had reached 1,484. The three communities had annual growth rates for 1983, 1984, and 1985 of less than 2%. In comparison, the population of the hamlet of Norman Wells jumped from 420 in 1981 to 678 in 1985, an increase of just over 60% in four years. In addition to these 'permanent' residents, Norman Wells also housed temporary workers who commuted by air from other northern communities and Edmonton.

These workers lived in self-contained work camps. At times, there were approximately 1000 to 1200 temporary workers living in Norman Wells. The residents of Norman Wells were subjected to a demographic impact not experienced by residents of the other three communities.

Income Impacts

During the construction of the Norman Wells Project, individual and family income rose sharply for residents of Norman Wells. For the other three communities, average incomes rose much less. For example, the median household income for Norman Wells residents was $38,000 in 1981/82 and $44,000 in 1983/84. In 1985, all households in the four communities were asked to indicate if they had gained economically from the Norman Wells Project. Nearly half of the households in Norman Wells indicated that they had gained either through jobs or business contracts. In the other three communities, much smaller percentages of the households indicated that they had gained from the Norman Wells Project (30% at Wrigley, 26% at Fort Norman, and 22% at Fort Simpson).

Income Differences

The explanation of income differences between the four communities involves a number of factors, including education levels. In 1985, the percentage of residents with a high school education was 73% in Norman Wells, 35% in Fort Simpson, 23% in Fort Norman, and 16% in Wrigley. Not surprisingly, median household incomes in each community correspond to educational levels. In 1982, the median income in Norman Wells was $38,000; in Fort Simpson, it was nearly $30,000; in Fort Norman, it was $21,000; and in Wrigley, it was only $8,000. Almost all of the adults without a high school diploma were Native Canadians. The income differences between Natives and non-Natives living in the four communities revealed that while income levels rose during the construction period, the gap in incomes between the two groups did not narrow. The median income for non-Natives in 1982 was $37,500 and for Natives $22,220, making a difference of $15,280 between the two groups. By 1985, the median incomes were $37,777 and $22,727 respectively, leaving a gap of $15,050.

Employment Impacts

Employment in the four communities increased during the Norman Wells Project. Most construction work took place at Norman Wells; in the other three communities local residents were employed by the pipeline contractors or by local firms holding pipeline-related contracts such as clearing the right of way. Many of these jobs and contracts were for a short period of time, often less than a month. For instance, for the year 1984/85, 57% of the workers at Norman Wells had held their present job for less than one year. During the four-year construction period, employment levels peaked in 1984 at nearly 1500 employees. By 1985, the size of the labour force had decreased to levels only slightly higher than those existing before the Norman Wells Project. The demand of large-scale construction projects for labour demonstrates a serious problem for northern development. The demand quickly outstrips the capacity of the local labour market to supply workers, thereby shifting recruitment to southern labour sources. Added to this problem is the limited number of skilled workers in the northern labour force. For these two reasons, many jobs are filled by workers from southern Canada. Many in the three Native communities were disappointed with the number of jobs obtained and this disappointment was fuelled by unrealistic expectations: 'As the first winter of construction drew near, everything seemed to work together to convince everyone in a certain part of the Mackenzie Valley that they were going to get a job' (DePape and Cairns 1985: 20).

In the final year of the construction project, unemployment rates in each community rose. At Norman Wells, the percentage of unemployed rose from zero in 1982 to 2.7% in 1985, at Fort Norman, from 14.7% to 23.5%, at Fort Simpson 10.6% to 19.7%, and at Wrigley 15.4% to 20.6% (Green and Stewart 1986: 6). Except for Norman Wells where there was a small in-migration of workers, the explanation for the increase in unemployment figures is that the construction project encouraged more Native adults to seek jobs. With the end of the Norman Wells Project, the number of jobs available dropped to pre-construction levels while the number seeking employment had increased well beyond the 1982 level. This finding supports the notion that there is considerable hidden unemployment in Native communities, i.e., many people wish to work but are not actively seeking employment.

Local residents also provided their opinions about the need for more jobs in their communities. As expected, the perceived need for more jobs was lowest in Norman Wells where employment rates are high. In 1982, for example, only 45% of the residents of Norman Wells felt there was a need for more jobs in their communities, compared to 94% in Fort Norman, 76% in Wrigley, and 89% in Fort Simpson. What was not expected was the shift in perception about the need for jobs by 1985. Sixty-four per cent of the residents of Norman Wells felt there was a need for more jobs in their community, while everyone in Fort Norman and Wrigley perceived a need for more jobs. In Fort Simpson, the 1985 figure had increased to 94%. These figures demonstrate the desire for more wage employment in Native communities, indicating the need for more cash income than can be produced by traditional pursuits.

Impact of Air Commuting

The local labour force at Norman Wells was far too small to satisfy the demand for construction workers at the oilfield. To obtain a skilled workforce, Esso employed two air-commuting systems. The southern system connected Edmonton with Norman Wells while its northern counterpart linked over a dozen communities in the Northwest Territories with Norman Wells.

Under normal market conditions, Esso would have obtained all of its construction workers from southern Canada, using Edmonton as its main source of labour. These employees would be shuttled to and from Norman Wells on two Boeing 727 jets. While their work schedules varied from one to four weeks in and one week out, air transportation costs would be minimized by using large fully-occupied aircraft. In this way, transportation economies of scale would be realized. But the Norman Wells Project was not left entirely to the market place. The federal government intervened at the project assessment stage, insisting that Esso make every effort to employ northerners. Federal reasoning was based on the premise that more wage employment in the North would lead to community and regional development and reduce the high numbers of unemployed Native adults. Esso responded by creating a second air-commuting system for residents of the territories. Northern rotational workers utilized the scheduled air line routes found in the Mackenzie Basin.

In the course of the project, there were over 1700 rotational workers. While a few were employed for the entire four-year construction period, the nature of construction work dictated that most worked for short periods of time, often less than six months. Almost two-thirds of these workers lived north of the 60th Parallel. Most came from communities in the Northwest Territories (Yellowknife had 288, Fort Smith 113, Hay River 103, Inuvik 103, and Fort Good Hope 91). The remaining southern air commuters formed 37% of the rotational workforce. The demand for labour by Esso varied, reaching a peak in 1984. This varying demand is reflected in the annual number of rotational workers. There were 261 shift workers in 1982, 511 in 1983, 836 in 1984, and 138 in 1985. These figures indicate the short-term nature of construction employment, the variation of demand during the construction period, and the ability of the northern labour force to increase its share of the Norman Wells job market. For example, nearly 540 native rotational workers were employed over the four-year period. All but 12 were from communities in the Northwest Territories. The northern air-commuting system played a key role in spreading the job benefits to communities beyond Norman Wells. Fort Good Hope had the largest number of native shift workers (81), followed by Fort Norman (50), Fort Franklin (44), Fort Simpson (41), and Fort Providence (29). The distribution of native rotational workers

Table 6.2
Air-Commuting Workers: Norman Wells Project, 1982–85

Territory/Province	No.*	%	Native	%
Yukon	102	5.9	7	6.9
Northwest Territories	997	57.1	526	54.0
Alberta	636	36.4	5	0.8
Other Provinces	11	0.6	0	0.0
Total	1746	100.0	538	31.4

*The number of workers is the sum of the annual employment figures obtained from companies who responded to the Norman Wells Socio-Economic Impact Program's business survey. Fly-in workers may or may not have been employed for the whole year. For 33 of the 1746 rotational workers, no information was provided on their ethnic status.

SOURCE: Norman Wells Socio-Economic Impact Monitoring Program.

by NWT communities reveals a pattern controlled by the size of the workforce in each community and accessibility to Norman Wells by scheduled aircraft. Just over 60% of the rotational workers came from the four major centres of Yellowknife, Fort Smith, Hay River, and Inuvik. Yellowknife is the dominant urban centre while Fort Smith, Hay River, and Inuvik are important regional centres. All four had direct jet aircraft connections with Norman Wells.

Impact on Country Food

Native peoples remain strongly attached to the land, but the desire for jobs on the Norman Wells Project signalled another element of the Native mixed economy. The Native mixed economy, however, continues to exhibit strong ties to the land as indicated by the continued high consumption of country food by Native people during the construction period.

Increased job opportunities provided by the Norman Wells Project resulted in more cash income in the hands of Native families. If store food was preferred over country food, this increased cash would have allowed for the purchase of more. Yet there was little change in the consumption of country food over the four-year period. From 1982 to 1985, the average household consumption of country food remained virtually unchanged, increasing from 41.7% of all food consumed to 42.6%. While the Norman Wells Project represented an additional intrusion of the modern industrial economy into the lives of these Dene, it did not alter their habits of harvesting and eating country food. The persistent use of country food reconfirms its significance to Dene culture and their commitment to their land-based economy (Bone 1988).

Social Impacts

Social impacts of the Norman Wells Project were not easily identified. One reason is that people are reluctant to discuss personal difficulties, such as alcohol abuse or job lay-off. A second reason is that social impacts may not surface immediately. A third factor is that the vast bulk of the construction work took place either in Norman Wells or along the pipeline route. People in the three Native communities saw few signs of the Norman Wells Project. These three points help explain the community response patterns to the question: 'In social terms, what effect has the construction of the Norman Wells Project had on your community?' At Norman Wells, the majority saw the project as having very strong positive effects on its social order while the second largest group thought that the negative and positive impacts were about balanced. In sharp contrast, residents of all three communities with a large Native population were less positive about the social effect.

While Norman Wells residents generally felt that the social impacts of the Norman Wells Project were good for their community, those in the other three communities were less certain about the impact and many chose a neutral position of 'no effect'. A follow-up question attempted to identify particular social impacts by asking the respondents to rank the social impacts, whether negative or positive. In Norman Wells, the most common concern was 'more transients'–the presence of large numbers of construction workers–while the next one was an appreciation of improved community facilities and services. Here, residents were likely associating the expanded business sector, the new school, and improved roads with the Norman Wells Project. At both Fort Norman and Wrigley, the major social impact was job experience and training opportunities while increased alcohol and drug abuse was seen as the major shortcoming. At Fort Simpson, the number one concern was also increased alcohol and drug abuse.

The problem of substance abuse existed before the Norman Wells Project and is part of a larger social problem, that of adjusting to settlement living and the dominant society. While there is no direct evidence that Native workers spent their wages on alcohol or drugs, the strong response from residents indicates the seriousness of this social problem.

The Norman Wells Project was a major energy construction undertaking. It took just four years to complete. From industry's perspective, the Norman Wells Project was a successful one, proving that northern oil can be shipped to southern markets. Its socio-economic impacts indicated that most economic benefits accrued to Norman Wells, and that social problems were minimal because construction activities and workers were kept away from Native centres. Social problems such as high unemployment rates to alcohol abuse existed in northern communities but these problems could not be remedied by the Norman Wells Project.

Selected Readings

Bone, R.M., 1984. *The DIAND Norman Wells Socio-Economic Monitoring Program*. Report 9–84. Ottawa: Department of Indian Affairs and Northern Development.

Dacks, Gurston, 1981. 'The Economic Future: Non-Renewable Resources', in *A Choice of Futures: Politics in the Canadian North*. Toronto: Methuen.

Duffy, Patrick, 1981. *Norman Wells Oilfield Development and Pipeline Project: Report of the Environmental Assessment Panel*. Ottawa: Federal Environmental Assessment Review Office.

Green, Milford B. and David A. Stewart, 1986. *Community Profiles of Socio-Economic Change, 1982–1985*. Report 9–85. Ottawa: Department of Indian Affairs and Northern Development.

Stewart, David A. and Robert M. Bone, 1986. *Norman Wells Socio-Economic Monitoring Program: Summary Report*. Report 1–86. Ottawa: Department of Indian Affairs and Northern Development.

Chapter 7

ENVIRONMENTAL IMPACT OF RESOURCE PROJECTS

The industrial economy has been hard on the North. While distant from the industrial centres of the world, the North has not escaped from the environmental consequences of industrial complexes in the developed areas of the world. Pollution from these factories and cities has entered the global circulation system and some of its fallout has been deposited in Canada's North. More direct industrial impacts have been caused by resource industries and urban settlements in northern Canada.

The polar environment is particularly susceptible to environmental damage from local and global pollution. Since polar terrestrial and marine ecosystems receive little solar energy for biological processes, their life-forms live close to the margin of existence. Permafrost and ice-covered lakes and seas slow biological activities. A small amount of pollution can have dramatic impact and the capacity of the polar ecosystems to recover from such damage is extremely slow. The fragile northern environment is, therefore, more susceptible to environmental damage than are other natural ecological systems.

What form do these pollutants take? Some are air-borne particles while others are water-borne. Evidence of environmental damage in the North is not always obvious. Such damage includes acid rain where northern forests and lakes in eastern Canada have been adversely affected by emissions of sulphuric acid from industrial plants in southern Canada and the United States; the thinning of the ozone layer over the Arctic caused by the discharge of aerosols into the atmosphere; and the appearance of toxic chemicals in the arctic food chain which may have originated in industrial factories found in other parts of the northern hemisphere. Then there is the greenhouse effect, which suggests that global atmospheric pollution is warming the climate. Such a rise in surface temperatures would have a profound effect on the polar environment and destabilize the permafrost equilibrium. At a local level, pulp and paper mills have fouled northern waters by discharging toxic effluent, uranium mills have spilled radioactive waste, and oil upgraders have emitted sulphur dioxide into the atmosphere. The effects of both direct and indirect industrial contaminants on humans is not always clear, but the presence of toxic chemicals in the food chain is, at the very minimum, a danger sign.

The Resource Economy and the Environment

The resource economy is hard on any environment because of its extractive nature and toxic waste products. The legacy of past pollution is shown in Figure 7.1. Large-scale resource projects, such as operating open-pit mines, clear-cutting vast stretches of the boreal forest, and building hydroelectric dams, water reservoirs, and river diversions, have transformed the natural landscape and, in doing so, have seriously affected the ecological balance of nature within that region. While these environmental changes were acceptable, even welcomed, by Canadians in the past, they are so no longer. Public pressure on governments to lower acceptable levels of environmental damage and pollution has met with success. Environmental standards for resource industries are much more demanding than those existing prior to 1990. Environmental groups continue to challenge industry at every turn, pushing for zero pollution and no environmental dislocations. Society supports these environmental groups but it also wants the jobs and regional growth generated by the resource industry. The federal government is attempting to set standards that minimize harm to the environment and still encourage resource development. The resource industry accepts the new environmental regulations but argues that it needs both time and public money to refit old plants to meet the higher standards. Resource industrialists claim that 'too demanding' regulations will prevent mines and mills from remaining competitive in world markets, forcing local plant closures and the loss of northern jobs.

The trade-off between the northern environment and resource development is a critical political issue. In the North, the resource industry forms the basis of the territorial and northern provincial economies, but it also poses a threat to the environment. This threat is more acute in the polar ecological systems than in most other ecological systems because of their much slower biological regeneration and the widespread existence of

Figure 7.1
Pollution in the North

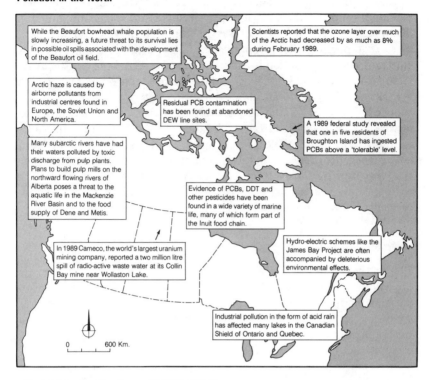

While the Beaufort bowhead whale population is slowly increasing, a future threat to its survival lies in possible oil spills associated with the development of the Beaufort oil field.

Scientists reported that the ozone layer over much of the Arctic had decreased by as much as 8% during February 1989.

Arctic haze is caused by airborne pollutants from industrial centres found in Europe, the Soviet Union and North America.

Residual PCB contamination has been found at abandoned DEW line sites.

A 1989 federal study revealed that one in five residents of Broughton Island has ingested PCBs above a 'tolerable' level.

Many subarctic rivers have had their waters polluted by toxic discharge from pulp plants. Plans to build pulp mills on the northward flowing rivers of Alberta poses a threat to the aquatic life in the Mackenzie River Basin and to the food supply of Dene and Metis.

Evidence of PCBs, DDT and other pesticides have been found in a wide variety of marine life, many of which form part of the Inuit food chain.

In 1989 Cameco, the world's largest uranium mining company, reported a two million litre spill of radio-active waste water at its Collin Bay mine near Wollaston Lake.

Hydro-electric schemes like the James Bay Project are often accompanied by deleterious environmental effects.

Industrial pollution in the form of acid rain has affected many lakes in the Canadian Shield of Ontario and Quebec.

0   600 Km.

Adapted from *Canadian Geographic* 1990: 112–13.

permafrost. For example, toxic discharges into the rivers and lakes take much longer to disappear than in tropical or temperature climates. Likewise, construction of roads and pipelines in the permafrost zone can disturb the polar vegetative cover and initiate a process of erosion and subsidence that can last for decades. Hydroelectric projects, including the diversion of rivers, can disrupt regional ecological systems. For these reasons, the Arctic and Subarctic regimes of the North may be described as 'industrially vulnerable' environments. Unfortunately, only in the last decade or so has society become aware of the dangers and hidden costs of uncontrolled industrial development. Pressure from society and, in turn, from governments is forcing the resource industry to reduce harmful impacts on the environment and to pay for more expensive but less polluting technologies. Nowhere is this change in attitude toward the environment more apparent than in northern Alberta where proposed pulp plants have had to reassess their plans in order to reduce the discharge of toxic pollutants into northern flowing rivers.

The popular image of the North as a region unaffected by industrial wastes is disappearing quickly as more and more reports of arctic pollution appear in the press, on television, and in scientific journals. It is now evident that the northern environment has received toxic pollutants from global industries. This pollution ranges from relatively benign dust particles to radioactive fallout from nuclear bomb testing, although the most common form of pollution is acid rain, caused by coal-burning factories. Acid rain has affected the lakes and forests of Ontario and Quebec; the use of chlorofluorocarbons in aerosol sprays, refrigerants, and foam has reduced the depth of the ozone layer over the Arctic; and air-borne pesticides, having drifted from other lands to the North, are found in the northern food chain. While it would be incorrect to suggest that the North is more polluted than southern Canada, it is equally incorrect to assume that the Arctic remains a pristine environment. Within the northern society, Native people suffer the most from industrial pollution because they rely heavily on local sources of food.

Emergence of the Environmental Movement

Before the 1960s, conservationists had focused their attention on the preservation of the natural environment and the creation of parks. As more and more examples of industrial pollution were reported in various countries, environmental issues gradually took on political importance. People in industrial countries realized that industrial and urban pollution was having an adverse effect on their lives, and support for the environmental movement grew. At the same time, scientists examining the sources of pollution discovered that chemicals previously thought to be benign were actually toxic and that such human-induced contaminants were found in food chains. Among these chemicals was chlorine, which is used extensively by the pulp and paper industry. Pesticides such as DDT were also found in these food chains. DDT was used extensively by farmers prior to 1971 when it was banned in Canada. Its chemical components are still found in the soil and water. Minute particles of residues of organochlorine compounds have been discovered in the Arctic food chain (Bidleman and others 1989). These contaminants had been transported to the remotest areas of the Arctic.

Other forms of pollution are less obvious but potentially more dangerous to human life. Local industrial activities can create dangerous conditions. A case in point occurred near Yellowknife in the 1960s, where arsenic levels in the town's water supply were found to exceed the permissible level of 0.05 parts per million about 15% of the time (Hare 1973: 260). The source of the arsenic was a gold refining process which involved roasting gold ores containing arsenopyrite. The process created an arsenical dust, much of which settled near the townsite of Yellowknife. Given enough accumulation in humans, other industrial substances such as DDT and PCB can become extremely toxic and eventually poisonous. Since both contaminants are found in the Arctic food chain, and since Inuit people eat large amounts of seal, caribou, and whale, there is a potential health risk to the Inuit.

Although many industrial pollutants are invisible, Arctic Haze and ice-fog are very noticeable examples of pollution. Arctic Haze is a form of smog caused by coal-based industries in Europe and Asia, while ice-fog is a smog-like feature occurring in larger urban centres such as White-horse and Yellowknife (Vignette 7.1). Both features occur during pro-longed cold spells and severe surface inversions. Coal-burning factories in Japan, China, the Soviet Union, and Europe produce sulphate, carbon, and other pollutants that are transported by air currents to the Arctic.

Reaction to these negative aspects of industrial development mobilized such environmental groups as the Sierra Club in the United States and Pollution Probe in Canada. The Canadian Arctic Resources Committee forged a link with Native organizations in its efforts to reduce environmental damage in the North. These and other environmental groups challenged industry and governments to reduce all forms of industrial damage to the northern environment. Eventually, legislation was passed to regulate the discharge of toxic wastes into the environment. The federal government banned the use of DDT and has entered into an agreement with the United States to reduce acid rain. It has also passed stiffer environmental legislation. New pulp and paper mills, for example, must meet higher standards for toxic waste discharge. Owners of older mills are under pressure to upgrade their operations and some are claiming that the costs are too high for these mills to remain commercially viable. The resource industry is confronting more and more demanding environmental rules. Recent scientific advances allow instruments to detect minuscule amounts of toxic contaminants undetectable only a decade ago. This allows regulatory agencies to lower the levels of toxic waste permissible in industrial effluent. Also, scientific research has been able to identify more toxic substances. Dioxins, for example, were declared toxic by Environment Canada in 1990 and new regulatory controls were introduced under the Environmental Contaminants Act regulating chemicals that are harmful to health or environment. These scientific advances are likely to continue, and our society faces two related questions: (1) is there an acceptable level of industrial pollution? and (2) is there a point where jobs come before environmental concerns?

Environmental Agencies

In the early 1970s federal and provincial governments created environmental assessment agencies in a number of departments. The federal agency, the Federal Environment and Review Office (FEARO), was formed in 1973, and similar agencies were organized by provincial governments. Provincial environmental units are responsible for assessing pro-

---

Vignette 7.1 Ice-Fog in the Arctic

Visibility in the polar regions is generally clear. However, there are characteristic conditions that dramatically limit visibility near the ground. One feature unique to the polar regions is ice-fog, which forms when a continuous supply of water vapour is released into air with a temperature of $-30°C$ or below and condenses into tiny ice crystals. Such conditions are common in arctic towns in valleys, where an inversion causes low temperatures and restricts mixing of air, and where combustion associated with vehicles and heating plants contributes more water to the atmosphere than can be absorbed without condensing. Fairbanks is a notorious example; when temperatures remain below $-40°C$ for a week, visibility at ground level is reduced to less than 10m, although the fog may only be 10m thick (Benson 1969, 1970). More serious than the ice crystals themselves is the pollution associated with the fog. Lead and carbon dioxide concentrations exceed those found in any other urban centre on Earth. Ice fogs may occur in Fairbanks any time from late November to the end of March. They are a growing problem in smaller Canadian Arctic towns such as Whitehorse, Inuvik, and Frobisher Bay, and are well known in eastern Siberian towns.

SOURCE: Sugden 1982: 60.

---

posed projects in their respective provinces, while their federal counterpart (FEARO) has sole jurisdiction in the two territories. If federal interests are involved in a proposed industrial project located in a province, then FEARO assesses the proposal. For this reason, FEARO reviewed the Nuclear Power Station at Point Lepreau, New Brunswick in 1975, and the Banff Highway Project, Alberta in 1982. A court decision in 1989 called for the federal government to conduct a review of the Rafferty Alameda Project in Saskatchewan even though its provincial counterpart had already approved the project. This court ruling suggests that Ottawa will play a more active role in assessing projects in the provinces.

The federal Environmental Assessment and Review (EARP) consists of three basic stages: screening, initial environmental evaluation, and formal review. Once a project is submitted to Ottawa for consideration, it undergoes an environmental screening by the appropriate federal department. The screening procedure has three possible outcomes. These are: (1) the project can proceed without further environmental review other than compliance with existing policies and standards; (2) if the potential adverse environmental effects are not fully known, then a more detailed assessment (Initial Environmental Evaluation) takes place; and (3) if the potential adverse effects are considered significant, then a Formal Review is undertaken. The Initial Environmental Evaluation involves a detailed examination of the proposal industrial project. If this evaluation finds that the potential environmental consequences are not expected to be significant, then the project is approved. If this is not the case, then the project must be withdrawn or undergo a Formal Review. The Formal Review is carried out by an Environmental Assessment Panel appointed by the Executive Chairman of FEARO in consultation with the proponent of the project and other public agencies. FEARO provides the panel chairman, who is also a member of the panel, and a secretary. The panel issues a set of guidelines to the proponent of the project. The proponent prepares an Environmental Impact Statement (EIS). This statement is reviewed by the panel. The panel also holds public meetings to allow individuals and groups to express their views and to hear the views of others. When these hearings are completed, the panel prepares a report, recommending whether the project is environmentally accept-able and if so, under what conditions. The final decision is made by the Minister of Environment and the Minister of the department receiving the proposal. In cases of disagreement, the Federal Cabinet makes the decision. The Federal Environmental Assessment and Review Process has established seven review panels since 1975.

Under this review process, proponents of proposed projects prepare the environmental impact statement. This approach has been criticized because it allows the developer, rather than an independent body, to prepare the statement. Another problem is the difficulty in determining the acceptable limits of changes to the environment brought about by a proposed industrial project. In other words, these environmental agencies are not attempting to halt industrial projects but rather to ensure that damage to the environment is minimized. An example of this approach is provided by the Windy Craggy Project. Reportedly an extremely rich deposit of copper, this proposed open-pit mining project in northwestern British Columbia has been targeted as high-risk by environmentalists and environmental officials. From the company's perspective, an open-pit mine is most cost-efficient. However, the risks posed to the local water table and streams by the sulphur-rich tailings are considered too high. This reaction caused the company to withdraw its initial proposal and to submit a more costly but environmentally sensitive plan. The revised proposal has not, however, satisfied a number of environmental groups (Vignette 7.2).

In 1981, the Federal Environmental Assessment and Review Process began a four-year review of the Beaufort Sea Hydrocarbon Production and Transportation proposal submitted by Dome Petroleum Ltd, Esso Resources Canada Ltd, and Gulf Canada Resources Inc. The Beaufort Sea Environmental Assessment Panel was charged with identifying the project's major effects on the human and natural environment and recom-mending ways and means of dealing with these impacts (Tener 1984: 13). The Panel concluded that oil production at a rate of 15,000 cubic metres/ day (nearly four times greater than production at Norman Wells) and several small-diameter pipelines (400 mm) are acceptable on environmental and socio-economic grounds providing that certain conditions are fulfilled (Tener 1984: 26, 27, 102). These conditions include minimizing adverse social effects on, and maximizing lasting benefits to, northern people; creating an effective voice for local residents in monitoring and managing problems that may come with changes to their way of life; and establishing oil-spill response and clean-up capability in advance of oil production. Some ten years later, construction of the project has not yet begun because the world price of oil is too low.

Vignette 7.2   The Windy Craggy Mine Proposal

The proposed copper mine is located in the mountainous corner of northwestern British Columbia near Alaska, 80 kilometres downstream from Glacier Bay National Park and Preserve and 32 kilometres from the Kluane National Park. Geddes Resources Limited, a Toronto-based mining company, applied to the British Columbia environmental agency for permission to proceed with development. The original submission was unacceptable and the company withdrew it. Major flaws in the initial proposal were the danger of highly acidic water draining from the sulphide-rich tailings and the open-pit mine into ground water and local streams, including the pristine Tatshenshini River. The new mine plan, released in January 1991, is estimated to increase development costs by $100 million, making the total cost around $600 million. It proposes to accomplish three goals: (1) reduce the amount of waste rock by 50%; (2) separate sulphide-bearing rock from other waste rock (which will still be dumped on the nearby glacier), and (3) place the sulphide-bearing waste rock in a tailings empoundment. The revised proposal has not satisfied environmental groups such as the Western Canada Wilderness Committee, World Wildlife Fund, Sierra Club, National Audubon Society, and Tatshenshini Wild.

SOURCES: Environment Canada 1990, Geddes Resources Ltd 1990, Robinson 1991: B1, and Searle 1991.

The Mackenzie Valley Pipeline Inquiry

Without a doubt, the Mackenzie Valley Pipeline Inquiry, headed by Chief Justice Thomas Berger, changed the course of northern development. It challenged the claims of large companies, and influenced future environmental inquiries by ensuring that they included a strong social component and a place for public participation. But perhaps most importantly the inquiry was able to reach the Canadian public and change their attitudes toward mega-projects, the northern environment, and Native issues. Berger accomplished this feat by expanding the focus of the inquiry from one dealing with a pipeline to a much broader examination of the environment and of Native people, and by ensuring ample coverage by the media. This allowed Native leaders such as James Wah Shee, President of the Indian Brotherhood, to present their views on land claims and other issues directly to the Canadian public. For these reasons, the Berger

Inquiry was more than an investigation into a single project; it was an examination of northern development and the place of Native people in that development.

The proposed Mackenzie Valley Pipeline Project called for the building of a pipeline to bring natural gas from the Prudhoe Bay field through the Mackenzie Valley to markets in southern Canada and the United States. Two companies, Canadian Arctic Gas Pipeline Limited (Arctic Gas) and Foothills Pipe Lines Ltd (Foothills), applied to build this pipeline. Arctic Gas was a consortium of Canadian and American petroleum and pipeline companies while Foothills represented Alberta Gas Trunk Line and West Coast Transmissions. The estimated cost of the Arctic Gas proposal within Canada was $8 billion (Berger 1977: 16). During the inquiry, three key environmental and social issues surfaced. These were: (1) the problem of burying a gas pipeline in areas with permafrost, (2) the threat to the arctic environment along the Yukon coast, and (3) the potential disruption of Native life and the impact on Native traditional culture.

The report of the Berger Inquiry recognized that the environmental soundness of burying the proposed Mackenzie Valley pipeline in permanently frozen ground was uncertain (Williams 1979: 58–71). The multitude of potential problems presented by permafrost was one factor in the rejection of the project. The report was equally concerned that the gas pipeline along Yukon's arctic coast would pass through the calving area for the Porcupine caribou herd, and affect the animals during the calving phase of their life cycle. Berger (1977: xii) concluded that 'the preservation of the herd is incompatible with the building of a gas pipeline and the establishment of an energy corridor through its calving ground.'

The Berger Inquiry set the standard for future inquiries. First of all, industrial proposals are more carefully scrutinized by the federal environmental assessment and review process. Secondly, the role of public participation has been enlarged and hearings are held in the communities affected by the project. Lastly, environmental issues affecting Native people have become a critical area of assessment. The importance of social impact assessment (SIA) in the North lies in the fact that environmental changes caused by industrial projects have a particularly harsh effect on people dependent on hunting, trapping, and fishing. Northern pipeline proposals did not end with the Berger Report. In 1977, Foothills proposed an alternative route for Alaskan gas, the Alaska Highway Gas Pipeline Project. This project was approved in 1982, but has not yet been built because the estimated price of its gas in the Chicago market is still too high. Next in line was a joint project by Esso Resources Canada Ltd and Interprovincial Pipe Lines (N.W.) Ltd. Their proposal called for the expansion of oilfield production at Norman Wells and the construction of a 'small' 324-mm diameter (12-inch) pipeline in the Mackenzie Valley corridor to Zama, Alberta. The oil industry saw this project as a test case for northern pipelines and, if successful, a model for the much larger

Beaufort Sea oil and gas project. The Norman Wells Project (described in Chapter 6) was approved by a Review Panel in 1981, begun the following year, and completed in 1985.

The Norman Wells Inquiry followed in the footsteps of the earlier Berger Inquiry. Issues had not changed; the only differences were that Norman Wells was a smaller project, more limited in its geographic area than the one proposed by Arctic Gas, and built in the discontinous and sporadic zones of permafrost. Environmental concern focused on the complex problem of maintaining ground stability after placing the oil pipeline in permanently frozen ground. Unlike the Mackenzie Valley pipeline proposal, the trench to be dug for this pipeline was entirely in the zones of discontinuous and sporadic permafrost. Interprovincial Pipe Lines (N.W.) Ltd proposed to bury the pipeline at least one metre below the surface, in the active layer of the ground. The temperature regime of the pipeline was to be the same as the surrounding ground, i.e., at or below the freezing point. The company designed the pipeline for a maximum thaw settlement of 1 to 1.2 metres. Under those conditions, the integrity of the pipe should be maintained. At the hearings, concern was expressed about removing one metre of the peaty surface material during trenching for the pipeline (Crampton 1988: 24). Without this peaty surface insulation, high temperatures in the summer could warm the fill in the trench and thaw ice in the pipeline trench. The result could be considerable subsidence and a possible rupture in the pipeline. Another concern was that the light oil from the Norman Wells deposit could convey heat from unfrozen to frozen ground and therefore melt any ice remaining in the trench. At the Environmental Assessment Review Panel in 1980, the proponents assured the panel that ground subsidences would be limited to 1 to 1.2 metres over the twenty-year life span of the pipeline. At the end of the first year of operations, Williams (1986: 94–6) reported that the initial subsidence was much greater than anticipated. More than a metre of subsidence had occurred at many places and the company responded by filling in these depressions. Fortunately, no ruptures in the pipe had occurred by 1990. Accurate predictions of future subsidence in the pipeline trench are not possible because of the limited knowledge of the complex thermal behaviour of the ground surface. The Norman Wells Review Panel described the company's analysis of the thermal regime of the ground as 'cursory' and 'insufficient' (Duffy 1981: 33–4).

Past Environmental Disasters

Most of the burden of pollution and past damage to the northern environment will be borne by Native Canadians who rely heavily on the land and water for food. Industries have caused most of this pollution. In the past, untreated wastes were discharged into water bodies, an approach not unique to the North but a common feature of early industrial development

Vignette 7.3 Integrity of the Norman Wells Pipeline

Scientists at the Norman Wells Pipeline hearings were concerned about thermal conduction, warming of the disturbed surface of the pipeline route, and global warming. The first concern was that the moving oil could be expected to conduct heat or cold through the steel pipe and thereby affect the surrounding ground. Two possible problems were identified. One was that the oil could be cooled while flowing through a permanently frozen section of the pipeline route and then cause freezing of ground in an unfrozen section. If the previously unfrozen section contained large amounts of moisture, then the ground and the pipe could be subject to considerable ground heaving. In another scenario, areas of frozen ground surrounding the pipe could be warmed and thawed by the flow of oil. Again, should there be ice-rich ground beneath the pipe, then considerable subsidence could occur, possibly leading to the rupture of the pipe.

A second problem might arise through the removal of ground-cover. In the course of building the pipeline, the forested vegetation was removed to form a right-of-way, a trench was dug and then refilled. The surface of this exposed right-of-way would receive more solar energy, causing a much greater thawing. Repeated freezing and thawing of the ground around the pipe over a decade or so could lift the pipe out of the ground.

Lastly, the scientists testifying at the hearing argued that the greenhouse effect could result in the warming of the northern hemisphere by as much as 3 C° over the project life span of the pipeline.

SOURCE: Duffy 1981: 33–4.

in the world. Once society realized that such pollution was affecting the quality of life, pressure was placed on governments and industry to reduce industrial damage to the environment. With the various regulatory agencies and environmental groups now in place, most industrial projects built 20 years ago in Canada would not be acceptable today. Resource industries are particularly concerned because their past fouling of the environment has made the Canadian public wary of their activities. Now these industries are much more careful about their impact on the environment.

One of the worst examples of industrial pollution occurred thirty years ago and still affects the Ojibway people living along the English-Wabigoon River near Kenora, Ontario (Shkilnyk 1985). Between March 1962 and October 1975 Dryden Chemicals Ltd, a subsidiary of Reid Paper Ltd, produced chlorine and other chemicals used as bleach in the pulp and paper mill of Reid Paper. The mill flushed its waste products into the Wabigoon River. The mill effluent contained a relatively high level of mercury which worked its way into the aquatic food chain of the river system. In 1970, the Ontario Government discovered that the level of mercury found in fish in a 300–mile stretch downstream from the pulp and paper mill was dangerous to health, and advised the Ojibway community at Grassy Narrows and Whitedog reserves not to eat fish from these rivers. It also banned commercial fishing on all lakes and tributaries of the English and Wabigoon rivers. The impact on the Ojibway was staggering. Economically, they lost their commercial fishing and guiding income. Psychologically, they could no longer trust their environment because their traditional source of food, fish, was the source of their illness.

Large-scale hydroelectric projects have had a 'regional' impact on river basins. They have flooded vast areas of land, altered stream flows, caused shoreline erosion and deposition, and increased the mercury content of the water. In the past, the cost of pre-flood clearing of trees was considered too expensive, and anyone who has visited a reservoir is struck by shoreline slumping and vegetation debris. Unfortunately, nature does not correct such man-made problems quickly. Some 50 years after the construction of the Island Falls dam on the Churchill River in Saskatchewan, there are standing snags (dead trees) in its reservoir (Sokatisewin Lake) while along its newly formed shoreline, trees have a tilted appearance (Kabzems and Bernier 1975: 63).

Major hydroelectric projects have altered the physical character of river basins in British Columbia, Manitoba, and Quebec. In the case of Manitoba, hydroelectric development is an important element of its economy. The purpose of energy development is to supply low-cost electrical power to industrial users and to export surplus power to markets in neighbouring provinces and the United States. Initial plans called for the construction of hydroelectric projects on the Churchill and Nelson rivers. In 1961, the Kelsey hydroelectric dam was built and most of its power was consumed by the nickel smelter at Thompson. Five other dams have been constructed: Grand Rapids, Jenpe, Kettle Rapids, Long Spruce, and Limestone, and the Conawapa Dam proposal is currently undergoing an environmental assessment.

In the early 1970s, Manitoba decided to increase the water flowing through the Nelson River by diverting most of the flow of the Churchill River. A control dam was built at the outlet of Southern Indian Lake to prevent water from continuing to flow to the mouth of the Churchill River, and a channel was dug between the southern edge of the lake and the Rat River which flows into the Nelson River. The water level on Southern Indian Lake rose by 3 metres and in 1976 water began to flow into the Nelson Basin.

While the amount of electric power generated by existing power stations on the Nelson River increased, the diversion of water from the Churchill River to the Nelson River is viewed by environmentalists as an ecological disaster. Submerged vegetation still chokes the lake, causing oxygen depletion during the winter (Rosenberg, Bodaly, Hecky and Newbury 1987: 81). Mercury levels in fish have risen, making frequent consumption unwise. Changes in the lake waters have greatly diminished the whitefish population, resulting in the collapse of the whitefish commercial fishery. Commercial fishermen, most of whom are of Indian ancestry, have received compensation from Manitoba Hydro through the Northern Flood Agreement or, in the case of those fishermen living in South Indian Lake, from special compensation agreements with Manitoba Hydro (Rosenberg, Bodaly, Hecky, and Newbury 1987: 83).

Global pollution, particularly toxic particles belonging to chlorine-containing organic compounds, has affected the food chain. Starting with single-celled algae, these toxic substances move up the food chain, becoming more concentrated with each step. The main sources of these pollutants are the pesticide known as DDT and the industrial insulating material referred to as PCBs. In 1989, a federal government official announced that a study of polychlorinated biphenyls (PCBs) in the Inuit country food diet found relatively high levels in the food chain. Furthermore, the study revealed that virtually all of the PCBs and related organochlorine contaminants found in southern Canada are also present in the Arctic but usually at much lower levels. Beluga whales in the St Lawrence River, for example, have PCB levels 25 times higher than those in the Eastern Arctic. For the Inuit whose diet consists partly of country food, the very presence of PCB in wildlife is disturbing and perhaps injurious to their health. Government officials stressed that they did not believe that there was any immediate danger to those eating country food and that these traditional foods have greater nutritional value than many of the prepared foods available in the local stores. In 1990, a study of Arctic beluga whales in the Western Arctic revealed extremely high levels of cadmium and mercury. While the federal government has not yet set guidelines for cadmium, the Health Protection Branch in Ottawa recommends that people not eat fish containing more than 0.5 parts mercury per million. Since the whales captured by Native hunters during the course of this study had levels of mercury well above the federal guideline, the consumption of these mammals by the Inuvialuit may be injurious to their health.

Most resource industries recognize the need to be more environmentally sensitive and have adjusted their effluent disposal programs to meet the new standards. Still, the basic problem remains. Mining firms, for example, face the problem of disposing of toxic waste products. Tailing ponds are one solution but these ponds can quickly fill. A recent example illustrates this point. Hudson's Bay Mining and Smelting operates a nickel-copper mine near Namew Lake in Manitoba, some two kilometres from the Saskatchewan border. The company applied to Manitoba's Clean Environment Commission for permission to pump mine wastes into Chocolate Lake. Treaty Indians living just across the border at Sturgeon Landing and Cumberland House are concerned about the impact on their drinking water. Since Chocolate Lake drains into Namew Lake from which the two bands obtain drinking water, there is the possibility that the effluent could affect the supply of potable water.

In May 1989, two northern Saskatchewan Indian bands requested a federal environmental review of another Manitoba nickel mine's plan. The Chiefs of the Cumberland House and Sturgeon Lake bands believe that a federal study is required because the Manitoba Clean Environment Commission may not sufficiently consider Saskatchewan interests or those of Indian peoples. Their request has been strengthened by the Federal Court ruling on the Rafferty and Alameda dam projects near Estevan in southern Saskatchewan, namely, that the federal government cannot rely on provincial environmental assessments of projects but must conduct its own inquiry.

Crown Corporations and the Environment

Crown corporations play a key role in the northern economy, particularly in the hydroelectric sector. In the last year or so, provincial power companies have proposed large-scale hydroelectric projects for the Peace River in British Columbia, the Conawapa dam on the Nelson River in Manitoba, and a series of dams on the Great Whale River in Quebec (the third phase of the James Bay Project). Of the three projects, the Great Whale River Project is the most likely to result in massive changes to the ecological system found in northern Quebec. Environmental impacts caused by damming the Grande River (the first phase of the James Bay Project) are well documented (Gill and Cooke 1975, Berkes 1982, Rosenberg, Bodaly, Hecky, and Newbury 1987; Cloutier 1987, and Gorrie 1990). The James Bay Energy Corporation was incorporated in 1971 for the specific purpose of developing the hydroelectric potential of the rivers on the Quebec side of James Bay. It is a subsidiary of Hydro Quebec. This corporation describes the James Bay Project as the symbol of Quebec's creative genius and it sees the hydroelectric project as 'development in accord with its environment' (James Bay Energy Corporation 1988: vii). The James Bay Energy Corporation formed an environment department whose task was to minimize the environmental damage caused by the construction of this massive hydroelectric project. The construction of the James Bay Project created four different types of aquatic systems: reservoirs, diversion zones, rivers with reduced flow downstream from diversion sites, and one river in which the flow was increased. The project also impacted on the surrounding environment, namely work camps and roads. According to this report (1988: 25), 'the remedial work in each of these habitats had the same goal: either to improve their biological stability and productivity or to facilitate access and improve hunting, fishing, and trapping conditions of native people.'

The major physical changes caused by the creation of the James Bay Project include the flooding of large areas and radical changes in seasonal stream flows. Other environmental changes are: increase in the mercury concentrations in these flooded waters caused by the biological process associated with the decay of drowned vegetation, particularly trees, and the appearance of mercury in the food chain, including fish, after empoundment; shoreline flooding and erosion which eliminates valuable animal and waterfowl habitat; and changes in the salinity and temperature of riverine estuaries and James Bay itself that could affect ice cover and the habitat for fish and mammals.

Evidence exists that beluga whales are affected by changes in the salinity and temperature of waters caused by hydroelectric dams. By increasing the salinity and temperature of sea water at the mouths of rivers, hydroelectric projects have reduced the habitat of beluga whales and eliminated preferred calving sites. The St Lawrence beluga whales have already felt this environmental impact and no longer congregate at the mouths of the Bersimis, Manicouagan, and Outardes rivers (all of which are dammed), and the dammed rivers flowing into James Bay could have the same impact on the estuarine habitat of the Hudson Bay beluga whales (Burnett et al. 1989: 24).

Similar impacts are expected from the next two phases of the James Bay Project. These phases include further development of La Grande River and the start of development of the Great Whale River. These hydroelectric developments will add to the massive transformation of landscape of northern Quebec and alter the coastal waters (Vignette 7.4).

Forest Industry and the Environment

The forest industry is one of the major employers in the Subarctic. Consequently local residents are often inclined to minimize the potential environmental damage caused by a pulp and paper mill. Provincial governments also support the mills and encourage the development of new ones because they are anxious to expand and diversify regional and provincial economies. On the other hand, many Canadians–particularly those downstream from such mills–are adamant that the effluents be reduced and toxic substances be eliminated. In the late 1980s the federal government introduced stiffer regulations affecting the discharge of untreated or toxic industrial wastes into the environment. The forest industry must now meet much higher environmental standards than previously set by the federal government. The critical question is, what level of pollution is society willing to accept?

Pulp and paper mills' past degradation of the environment was so severe that the industry has been singled out for stiffer regulations. In the late 1980s, for example, it was responsible for half of all the industrial wastes discharged into water bodies and 6% of those sent into the atmosphere (Sinclair 1990: 39). No industry is more vulnerable to public pressures than the forest industry, for it depends on Crown-owned timber leases for its raw material. The forest industry realized that it must be more environmentally sensitive or else it could be denied timber leases,

Vignette 7.4   The James Bay Project

The scope of the James Bay hydroelectric project is staggering. It involves the transformation of the northern landscape in Quebec and a new life style for the Indians and Inuit of the Subarctic. This project will eventually harness the energy of all the rivers flowing through 350,000 square km of northwestern Quebec, thereby producing up to 28,000 megawatts of electrical power. At least a dozen rivers will have their water diverted, leaving shrunken waterways with pools of water and dried-up riverbeds. The water is to be collected in vast reservoirs and directed toward generating stations where electric energy is or will be produced. This massive project involves some 20 rivers and an area about one-fifth of Quebec. Its four phases are:

(1) La Grande, Phase 1: This phase of the James Bay Project was completed in 1985 at a cost of about $16 billion. Three dams, reservoirs, and power stations were built on La Grande River. These installations are named LG2, LG3, and LG4. Five rivers were diverted to provide more water for La Grande River.

(2) La Grande, Phase 2: This phase is under way and completion date is 1996. It includes six dams, reservoirs, and power stations at LG1, Brisay, Eastmain 1 and 2, and Laforge 1 and 2.

(3) Great Whale, Phase 3: Located north of La Grande, this river basin is planned to house three power stations which will require the diversion of two rivers.

(4) The NBR Phase 4: This project involves three large rivers, the Nottaway, Broadback, and Rupert. Water from the Nottaway and Rupert rivers is to be diverted into the Broadback River where eight power stations are to be built.

particularly in the more fragile northern environments. Two critical issues face the forest industry: clear-cut logging practices, and the discharge of toxic chemicals into the environment by pulp and paper companies.

Clear-cut logging practices have been banned or restricted in many parts of the developed world. The argument for clear-cutting is economic efficiency. Forest industry officials maintain that sustainable, balanced development of forest land can best be achieved by clear-cutting because natural and artifical regeneration of the forest can then take place. Environmentalists disagree and argue that regeneration is best after selective cutting. Other environmental problems associated with clear-cutting occur in mountainous and hilly terrain where serious erosional and flooding problems may result. The sizes of clear-cut areas normally are limited. Yet, a vast area (reportedly over 2,500 square km) was cut near Kapuskasing in northern Ontario in 1990 (Mittelstaedt 1991: A3). This assessment, gleaned from satellite photography, was presented to the hearings of Ontario's Environmental Assessment Board on forestry practices by Forests For Tomorrow, a coalition of five environmental groups that includes

Table 7.1
Proposed or Expanded Pulp and Paper Mills in Alberta

Company	Cost ($ billions)	Timber Lease (000 km²)
Daishowa Canada Co. Ltd	$1.2	40.71
Procter and Gamble Cellulose Ltd	0.5	21.16
Weldwood of Canada Ltd	0.5	9.96
Alberta Newsprint Co. Ltd	0.7	3.77
Alberta Energy Co. Ltd	0.3	8.03
Alberta-Pacific Forest Industries Inc.	1.3	73.43
Total	4.5	153.06

SOURCE: Nikiforuk and Struzik 1989: 63.

the Wildlands League, the Federation of Ontario Naturalists, and the Sierra Club. Europeans, concerned about Canadian clear-cut logging, are urging the European Parliament to pass legislation boycotting Canadian forest products (forest exports to Europe reached $4.5 billion in 1990).

The economic attraction of clear-cutting often outweighs costs to the environment. Recent developments in Alberta illustrate this conflict. To the Alberta government, the proposal to build pulp and paper plants in northern Alberta represented an opportunity to diversify the oil-oriented provincial economy. Alberta jumped at the opportunity to lease large tracts of timber to companies proposing to build large-scale pulp and paper mills. It and the federal government also provided public funds for the larger projects. Alberta-Pacific Forest Industries Inc., for example, has received a $300 million loan, and public funds estimated at $75 million to pay for the access roads and rail lines to its proposed bleached-kraft pulp mill near the community of Athabasca. The major investors in new or expanded mills are shown in Table 7.1.

While there are undoubtedly benefits of economic diversification and new jobs for northerners in these projects, the northern environment will be affected, timber will be cut, and industrial wastes will be discharged into the air and water. This trade-off should mean that economic benefits far outweigh environmental losses. Interestingly, public concern in Alberta has forced the proponents of the Alberta-Pacific to change their milling operation to reduce the volume of toxic chemical residues of dioxins and furans to be discharged into the Athabasca River. These changes were made after the environmental review board recommended halting the project until more was known about the potential impact of effluents from the mill on the Athabasca, Slave, and Mackenzie rivers. The company has replaced one chemical agent, chlorine, which is a major source of dioxins and furans with another one, hydrogen peroxide, which is believed to produce much lower levels of dioxins and furans. The federal environmental agency approved the new proposal and in conjunction with the provincial government announced that it would force other pulp mills on the Athabasca River to lower the level of waste discharge containing dioxins and furans. These mills include Weldwood near Hinton, Millar Western and Alberta Newsprint both near Whitecourt, and Alberta Energy Company near Slave Lake. All of these mills discharge higher levels of toxic chemicals into the Athabasca River than the estimated figures for the proposed Alberta-Pacific mill. The government of the Northwest Territories continues to oppose the project, claiming that its residents will pay for the environmental costs and yet receive none of the economic benefits.

Global Pollution

Global pollution of the Arctic has affected the northern environment in many ways. In this discussion, three types of global pollution are examined. These are acid rain, the greenhouse effect, and fall-out of radioactive debris. Two forms of global pollution are associated with the burning of fossil fuels while the third was produced by nuclear bomb testing. Once the contaminants are airborne, winds can carry them thousands of kilometres from their source, depositing some on the earth's surface while others remain in the atmosphere.

The tundra vegetation cover contains large areas of lichens which derive their nutrients from salts dissolved in surface waters. In the 1950s and 1960s, series of nuclear tests in the interior of China and in the Soviet Union took place. The westerly circulation system transported the radioactive contaminants to northern Canada. As radioactive particles of Strontium 90 and Cesium 137 fell to the ground, they lodged in the surface waters and were absorbed by the lichens. Caribou and other herbivores browsed on the tundra vegetation and ingested large quantities of the radioactive particles. During the 1960s, scientists with the Department of National Health and Welfare conducted studies of this contamination and they concluded that caribou meat contained Cesium 137 levels as high as 35 microcuries per pound, and that Inuit adults had Cesium 137 levels 20 to 100 times those found in Canadians living in southern Canada (Hare 1973: 261). Since the ending of nuclear testing, the danger of radioactive material in the food chain has declined but, unfortunately, fallout from the 1986 Chernobyl nuclear disaster in the Ukraine, USSR added new radioactive particles to the upper atmosphere and minute quantities were deposited across the northern hemisphere (Gould 1988: 58).

The term acid rain describes the deposit of acidic pollutants, chiefly oxides of sulphur and nitrogen, on the earth's surface. The principal source of these air-borne pollutants is industrial and urban emission of sulphur dioxide (SO_2). Current annual emissions of SO_2 in Canada and the United States are nearly 30 million tonnes, most of which come from coal-fired thermal electrical power plants. In Canada, significant sources of SO_2 emissions are copper and nickel smelters and heavy oil upgrading plants. Once in the atmosphere, sulphur and nitrogen dioxides are easily absorbed in water vapour and converted into sulphuric and nitric acids respectively. When washed out of the air by rain, snow, or fog, this acidic precipitation sets off a chain of chemical and biological reactions. The effects of acid rain are most pronounced on rivers and lakes where an increase in the acidity of the water can have a devastating effect on aquatic life, particularly fish populations. Thousands of lakes in Canada have already lost their fish population due to water acidification and many more are threatened.

When acid rain falls, chemical and biological reactions vary according to the acid-reduction capability of the surface. This buffering effect is generally low in the North because of the high acidic levels normally found in soils and water bodies. Adding to this high level causes significant environmental changes. Acid rain affects both soils and vegetation. Extensive forest damage has occurred in eastern Canada and the United States. A solution to the problem of acid rain may be at hand. In March 1991, the Prime Minister of Canada and the President of the United States signed a treaty calling for the reduction of SO_2 emissions.

Ever-increasing air pollution may also lead to the warming of our climate. This hypothesis is based on the assumption that air pollution is causing the greenhouse effect, i.e., the capacity of the atmosphere to trap long-wave radiation from the earth's surface (Ripley 1987). Proponents of this theory argue from the fact that carbon dioxide and other gases released during the burning of fossil fuels are changing the chemical composition of the atmosphere, thereby increasing its capacity to retain more long-wave radiation given off by the earth's surface. They believe that sufficient amounts of carbon dioxide have already been released into the atmosphere to have an impact on world temperatures (Davies 1985). Since the early 1960s, for instance, concentrations of carbon dioxide in the atmosphere have increased by almost 10% (Maxwell 1987: 2). The fact that the six warmest years ever recorded occurred in the 1980s supports the warming argument but it does not provide conclusive evidence (Chenard 1990: 59). This slight rise in temperatures may be a temporary phenomenon but proponents of the greenhouse effect believe that it marks the beginning of a long-term rise in temperatures. Other scholars believe that minor flucuations in global temperatures are a normal occurrence. They attribute temperature fluctuations to variations in solar radiation, shifts in the earth's orbit, and changes in ocean currents. Such 'temporary' variations account for past climatic changes such as the 'Little Ice Age' which lasted for some 400 years, roughly between 1450 and 1850 (Vignette 7.5).

Whatever the cause, global warming would result in significant changes in the northern environment. The warming, if it occurs, would be much

Figure 7.2
Northern Areas Most Sensitive to Acid Rain

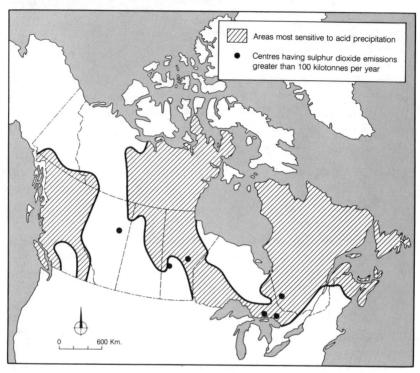

SOURCE: *The Canadian Encyclopedia* 1988: 12.

greater at higher latitudes than around the equator. Warming of northern lakes and seas would facilitate evaporation, causing more active convection currents in the atmosphere and more precipitation to fall in the northern hemisphere. It might melt glaciers and permanent ice pack over the Arctic Ocean, and thaw the permanently frozen ground. The release of such quantities of water could produce a rise in the sea level. Changes in temperature and precipitation would affect northern soils, vegetation, and biological life. One such change would see northern coniferous forest extending further north. Another would be to reduce the areal extent of permafrost. While a milder climate may be an advantage in some ways, the landscape underlain by permafrost will be subject to thawing and subsidence. The impact on highways, bridges, and pipelines could be devastating. Other possible changes are listed in Table 7.2.

As Canada's northern environment has changed from a natural landscape into an economic one, global pollution and resource development have damaged the environment. The North, as an integral part of the global ecological system, has been subject to global pollution generated

Table 7.2
Possible Effects of Global Warming on the North by 2075

Agriculture	— increase in agricultural potential
Natural Vegetation	— northward movement of the Subarctic
Permafrost	— increase in the depth of the active layer
	— increase in slumping
Periglacial	— reduction in extent of pattern ground
Snow and Ice	— duration of ice cover shortened
	— ice thickness reduced
	— reduction in size of glaciers
	— increase in iceberg calving

Adapted from Maxwell 1987.

Vignette 7.5   Climatic Change

How do present climatic conditions compare with those at various times in the past? A number of scholars have looked at this intriguing question, and Marsh (1987) places this question within a geological perspective: 'About two million years ago the climate cooled again, and the present Ice Age began. This period has been marked by several dramatic expansions and contractions of the world's ice volume. Over the past million years, global temperatures (equilibrium surface temperature) have fluctuated by about 4°. After the last glacial peak, about 16,000 years ago, climate grew warmer until 6000 years before the present (4000 B.C.). Since then climate has generally cooled, but the cooling trend has been interrupted by oscillations of about 2°C every 2000 years. Scientists generally agree that the earth is still in the last ice age, probably in a warm period between major glaciations, called an interglacial period.'

Climate change over the last thousand years is based on winter temperatures recorded in Europe. These records were often based on descriptions of crop failures rather than actual temperature recordings. From these data, it is clear that temperatures in the northern hemisphere declined around 1450 and began to increase after 1850. This 'cool' period is known as the Little Ice Age. Geological records of mountain glaciers independently confirm the timing and extent of the Little Ice Age.

SOURCE: Marsh 1987: 160.

by far-off industrial plants and agricultural operations. Acid rain is one visible sign of global pollution reaching into the Arctic. Resource development has affected the North too. Most of its impact has taken place in the Subarctic where forestry, mining, oil and gas, and hydroelectric developments are concentrated. To many Canadians, these environmental impacts are the price of progress; to northern Canadians, they are the hidden cost of development. Both agree, however, that past environmental disasters should not be repeated. Governments, responding to the demands of Canadians, have stiffened regulations governing the disposal of industrial wastes, taken a harder look at construction projects in permafrost areas, and are pumping money into reforestation projects. Environmental agencies such as FEARO assess proposed projects while others, using the Environmental Contaminants Act, regulate industry.

Governments are attempting to balance environment concerns and industrial demands. In the past, public environmental assessment agencies appeared to believe that economic benefits would outweigh environmental damage. Canadians today want a more environmentally responsible industry, one that ensures the continued well-being of the environment. Industry, both private and public, must shoulder the cost of reducing environmental impacts to acceptable levels, and governments must establish these levels. In this way the true costs of northern development can be determined and society is left with a more reasonable trade-off between economic benefits and environmental costs.

Selected Readings

Carson, Rachel, 1962. *Silent Spring*. Boston: Houghton Mifflin.

Chenard, M. Paul, 1990. 'Global Atmospheric Change', in *Proceedings of the 4th Conference on Toxic Substances*. Environment Canada. Ottawa: Minister of Supply and Services.

Davies, John A., 1985. 'Carbon Dioxide and Climate: A Review', *The Canadian Geographer* 29(1): 74–85.

Maxwell, B., 1987. 'Atmospheric and Climatic Change in the Canadian Arctic: Causes, Effects, and Impacts', *Northern Perspectives* 18(5): 2–6.

MacInnes, K.L., M.M. Burgess, D.G. Harry, and T.H.W. Baker, 1990. *Permafrost and Terrain Research and Monitoring: Norman Wells Pipeline*. Environmental Studies No. 64, Vol. 2. Ottawa: Minister of Supply and Services.

Rosenberg, D.M., R.A. Bodaly, R.E. Hecky, and R.W. Newbury, 1987. 'The Environmental Assessment of Hydroelectric Impoundments and Diversions in Canada', in *Canadian Aquatic Resources*, edited by M.C. Healey and R.R. Wallace. Ottawa: Department of Fisheries and Oceans: 71–104.

Saunders, Alan, ed., 1990. 'Pulp and Paper on the Athabasca: Economic Diversity vs. Environmental Disaster', *Northern Perspectives* 18(1): 1–12.

Searle, Rick, 1991. 'Journey to the Ice Age: Rafting the Tatshenshini, North America's Wildest and Most Endangered River', *Equinox* 55(1): 24–35.

Sinclair, William F., 1990. *Controlling Pollution from Canadian Pulp and Paper Manufacturers: A Federal Perspective*. Environment Canada. Ottawa: Minister of Supply and Services.

Tener, John S., 1984. *Beaufort Sea Hydrocarbon Production and Transportation Proposal: Final Report of the Environmental Assessment Panel*. Federal Environmental and Review Process. Ottawa: Minister of Supply and Services.

Williams, Peter J., 1986. *Pipelines & Permafrost: Science in a Cold Climate*. Ottawa: Carleton University Press.

III
NATIVE DEVELOPMENT

Chapter 8

ABORIGINAL POPULATION AND SOCIETY

Just over one-quarter of Canada's Aboriginal peoples live in northern Canada. They, like other Native Canadians, have a much more youthful population than other Canadians. They also have much shorter life expectancy. The demographic events associated with these and other population changes are described in this chapter. These events include the emergence of an extremely high natural rate of increase among northern Native Canadians and their settling into small urban centres. Since becoming urban dwellers, Native Canadians have been faced with making many adjustments to their ways. These adaptations may have changed Native family life and society. During the transition from a land-based people to an urban one, some cultural losses have occurred, and social problems have arisen. These cultural losses and social problems are not limited to the Canadian North but are found in other circumpolar countries.

When Columbus reached North America in 1492, northern Canada contained a relatively small population–probably fewer than 100,000 Native people. The exact size of this population is not known but Mooney (1928) and later Kroeber (1939) have suggested a figure of around 86,000 with about 60,000 inhabiting the Subarctic and the remainder occupying the Arctic. During pre-contact times, Indian and Inuit peoples depended on hunting, fishing, and gathering for their food supply. Since animal populations tend to follow a cyclic pattern, it seems plausible that the size of pre-contact populations would have declined in times of natural food shortages and increased in times of plenty. This hypothesis is based on the Malthusian principle that the size of human populations tends to increase (or decrease) as the food supply increases (or decreases). Other checks on population size involve social customs and destructive forces such as pestilence or famine. Major population declines among the pre-contact peoples of the North were likely associated with changes in food supply, such as the extinction of the woolley mammoth, or with climatic changes which would in turn affect animal populations. For example, it is believed the Little Ice Age reduced the presence of bowhead whales in the Arctic and thus limited the food supply of the Thule Eskimos.

In these hunting societies, the search for wild animals, fish, birds, and berries was a daily activity. Large kills, such as a whale or a number of caribou, could feed a large number for weeks, even months. At other times, the scarcity of game could lead to periodic–but localized–famines.

In the pre-contact period, the hunting strategy of northern peoples was based on mobility. Hunting groups usually consisted of extended families, which spent most of the year moving from area to area in harmony with the seasonal movements of animals. Their strategy was simple. The small size of the hunting group provided an efficient hunting unit for pursuing, killing, and processing game into food, shelter, and clothing. These tiny groups would reunite once or twice a year at sites where there was an abundant food supply, such as at the mouth of a river when fish were spawning. The nature of the hunting economy, whether practised by Indians or Inuit, required a high degree of mobility. A hunting party had to travel 'light', and rely on their skill to live off the land. Such a pattern of land use results in extremely low population densities and relatively few people.

Population Density and Carrying Capacity

After the disappearance of the Laurentide Ice Sheet, the technological level of hunting societies and the northern environment remained relatively stable. Prior to contact with Europeans, population density was largely controlled by local carrying capacity. The biological notion of carrying capacity suggests that a population cannot increase its size indefinitely and that there is a point at which a 'ceiling' is reached. Carrying capacity varies according to the nature of the resource base and the technological level of the society in question. A hunting society, for example, would have a low population density compared to the density of an agriculturally-based economy.

Across the 4.5 million square kilometres of Subarctic lands, pre-contact Indian bands had very low population densities. If the average density for the 4.5 million square kilometres of subarctic lands was one person per 100 km^2, then the Subarctic would have a population of 45,000. The

Arctic landscape presented an even less productive environment for terrestrial wildlife than the Subarctic but since the Thule Eskimos harvested much of their food from the sea, we can assume an average density of one person per 200 km² for 3.0 million km² of Arctic lands; its population would total around 15,000. In sharp contrast, the more populous and sedentary Iroquois tribes living in the more temperate St Lawrence River valley practised both agriculture and hunting. The carrying capacity of these southern lands permitted much higher population densities, perhaps as high as one person per 20 km². While these population estimates demonstrate varying carrying capacities of land for hunting societies, the population density figures are highly speculative because they are not derived from an analysis of potential wildlife harvesting but are arbitrarily selected. Mooney (1928), for example, suggested a pre-contact population of 26,000 in the Canadian Arctic and 60,000 in the Subarctic. Using his estimates, the population densities would be one person per 115 km² for the Arctic and one person per 75 km² in the Subarctic.

The concept of carrying capacity, while intuitively appealing, has a fundamental weakness–it ignores the capacity of human beings to adjust to new conditions, and to control their population size. Another weakness is that it assumes constant environmental conditions over time. The onset of the Little Ice Age represents such an environmental change. While our knowledge of past climatic conditions is hazy at best, the arctic climate was much cooler between 1450 and 1850. During that period the Thule whaling economy collapsed because much heavier ice cover impeded the movement of bowhead whales into arctic waters (McGhee 1984: 374). The new economy based on seals, caribou, and other game and fish supported fewer people–perhaps only half as many as before. For these reasons alone, population densities can provide only a rough indication of past populations. We must conclude that pre-contact populations in the Arctic and Subarctic were small but there is inconclusive evidence of exactly how small or how they have changed over time. This unsatisfactory situation leaves us with estimates by Mooney, Kroeber, and others (Jenness 1932, Driver 1961, and Helm 1981).

Early Historic Population Estimates

As Europeans began to occupy the fertile lands of North America, the original population was pushed aside and ravaged by diseases unknown before contact. Those in the North were more fortunate in that their lands did not attract white settlers. But like the southern Indians, they suffered from epidemics. By the turn of the twentieth century, the number of Indians was less than half of their pre-contact population.

The magnitude of this population decline can only be estimated by comparing the very rough pre-contact figures with more recent ones. In 1857, the first comprehensive estimate of the Indian and Inuit population

Vignette 8.1 Historic Estimates of Population Size and Density

An early estimate of population density and the carrying capacity of lands occupied by east coast Indians was made in 1611, by a Jesuit missionary. He reported that an area of some 415,000 km² was inhabited by approximately 10,000 Indians belonging to the Micmac, Malecite, Abenaki, and Montagnais tribes. Their hunting grounds constituted today's provinces of Nova Scotia, Prince Edward Island, and New Brunswick, the Eastern Townships of Quebec, and the states of Maine, Vermont, New Hampshire, Massachusetts, and Connecticut. While population densities varied from place to place, the average was one person per 41 km². Unfortunately, comparable figures for northern tribes were not made at that time, though such density figures should be much lower because wildlife would have been less plentiful. Nearly 300 years later, the 1881 Census of Canada provided an 1871 population density figure of one person per 342 km² for the Naskapi who continued to hunt caribou in the vast interior of Quebec and Labrador. The Naskapi had been involved in the fur trade during these 300 years. This involvement plus the impact of European diseases may have resulted in a smaller population (and density) than existed prior to John Cabot's discovery of Newfoundland.

SOURCE: Census of Canada 1880–81, Vol. IV: lvi-lxvi.

of British North America was made by Sir George Simpson for the Special Committee of the British House of Commons on the affairs of the Hudson's Bay Company (Census of Canada 1876, Vol. IV: lxxiv). Simpson's figure of nearly 139,000 was based on reports made by local Hudson's Bay Company officials on the number of Indians trading at their posts. Approximately 61,000 Indians were reported as trading at posts in the Subarctic. Added to these numbers was Simpson's guess that 4,000 Eskimos occupied the Arctic.

From pre-contact times to 1857, the population decline of Aboriginal peoples living in Canada may have dropped by about one-quarter, from 86,000 to 65,000. Nearly a quarter century later, the Census of Canada published a figure of around 108,000 Indians, Métis, and Inuit residing in Canada. This suggests that the Native population continued to decline during the last half of the nineteenth century. Much of this decline occurred in the West and was due to the demise of the buffalo herds and the displacement of Métis and Indians by agricultural settlement. Since northern Native people were not directly affected by agricultural settlers, their population may not have declined so seriously. This hypothesis is supported by the population estimates of Mooney and Simpson, who calculated that in pre-contact times 86,000 out of 220,000 people (39%) lived in the North, while in 1857 65,000 out of 134,000 (49%) were in the North.

In 1871, the Census of Canada reported a Native population of 102,358. By the next census, this figure had increased to 108,547. Yet, the difficulty of enumerating nomadic Indians on Census Day remained a challenge

Table 8.1
Native Population in Canada, Pre-contact to 2001

Year	Number	Notes
Pre-contact	220,000	Mooney's estimate
1857	139,000	British North America included the Oregon Territories, part of which was ceded to the United States. Approximately 5,000 Indians were reported in these U.S. lands.
1871	102,385	
1881	108,547	
1901	127,941	Half Breeds were identified for the first time. Their population was reported as 34,481 while the number of Indians was 93,460.
1911	105,611	
1921	113,724	
1931	128,890	
1941	125,521	The 1941 census excluded Half Breeds.
1951	165,607	
1961	220,121	
1971	312,760	
1981	491,500	Recorded only single responses.
1986	711,720	Recorded multiple responses. Also, 136 reserves did not participate in the 1986 census enumeration and Statistics Canada estimates that the number of Indians not enumerated totalled 44,733.
1991	958,452	Figures for 1991 and 2001 were estimated by Hagey, Larocque, and McBride 1989.
2001	1,145,109	Estimated.

SOURCES: Censuses of Canada and Hagey, Larocque, and McBride 1989 I:22.

and estimates were often used. The estimated Native population in the territories in 1881 was 49,472; the distribution of Native Canadians by provinces is shown in Table 8.2.

While little is known about these Native populations prior to the 1881 census, Table 8.1 attempts to illustrate general population trends for Canada, and by implication the North, from pre-contact to the year 2001. It indicates three major demographic periods. These are a relatively stable pre-contact population ranging between 200,000 and 250,000; during the initial contact a substantial drop in population, leaving the Native population at around a hundred thousand and a long period of little or no population growth largely because periodic epidemics nullified any natural growth; the last period, associated with rapid growth, is expected to continue into the next century.

The Impact of Diseases

The single most important factor in population loss in the North was disease. Contact with Europeans and later Euro-Canadians led to the

Table 8.2
Indian Population by Provinces and Territories, 1881

Political Units	Indian Population	Total Population	Per cent Indians
Prince Edward Island	281	108,891	0.3
Nova Scotia	2,125	440,572	0.5
New Brunswick	1,401	321,233	0.4
Quebec	7,515	1,359,027	0.6
Ontario	15,325	1,923,228	0.8
Manitoba	6,767	65,954	10.3
British Columbia	25,661	49,459	51.9
The Territories	49,472	56,446	87.6
Total	108,547	4,324,810	2.5

SOURCE: Census of Canada 1880–81, Vol. I, Tables I and III.

spread of previously unknown diseases through North America. Aboriginal peoples had little resistance to these new diseases and many succumbed. Extremely high death rates occurred during epidemics, resulting in sharp declines in local population. Until medical services were readily available in the Canadian North in the 1950s, among the illnesses ravaging the Native population were diphtheria, influenza, measles, smallpox, typhoid, and tuberculosis. Because of the remote nature of the North in early days, the diffusion of contagious diseases was a slow process. In the early nineteenth century, it might take a decade or more for a disease like smallpox to spread from Fort Edmonton to the shores of the Beaufort Sea where the Mackenzie Delta Eskimos lived. In the Arctic, the more isolated Inuit were not involved in the fur trade in a substantial way prior to the twentieth century. However, they did contact diseases from whalers, and epidemics of measles, influenza, and smallpox struck suddenly and without warning.

Within the Subarctic, other threats came from periodic food shortages and fighting between tribes. Food shortages were often the result of earlier excessive trapping/hunting. Tribal fighting involving European weaponry often resulted in high casualties. In 1823, the Dogrib Indians attack the Yellowknife Indians, killing most of the group. The Yellowknife Indians did not recover their former numbers and by the 1960s, they were no longer an identifiable people.

Beyond these generalities about population losses suffered by tribes following contact with Europeans, virtually nothing is known about their demographics, particularly vital rates. What is known is that diseases could decimate a tribe.

In 1535 in an early phase of European exploration, Jacques Cartier, reached Stadacona and Hochelaga, settlements of the St Lawrence Iroquois. Later in the century, these people abandoned their settlements.

Their disappearance is a puzzle. There are a number of possibilities. One is that there was tribal warfare, over either hunting and fishing territories or access to European fishermen with whom trade could be conducted. According to this hypothesis, the Iroquois were driven away. Another possibility is that Cartier's men infected the Iroquois with smallpox or some other contagious disease, and the remaining Iroquois might have fled the region.

The history of Newfoundland's Beothuks acknowledges disease as one of the factors leading to their demise. During most of the fifteenth century, the Beothuk, who may have numbered around one thousand, were in contact with Basque fishermen. The fishermen came ashore to dry their fish, obtain fresh water, and trade with the Beothuk. Contacts between the two groups were limited because the Basque sailed back to Europe each fall. Conflicts may have resulted through Indians taking various items from the shacks built by the fishermen, and by fishermen taking Beothuk women. These clashes caused the Beothuks to retreat to the interior of Newfoundland. Contact was renewed in the eighteenth century when permanent coastal settlements were established. Since both New-foundlanders and Beothuk engaged in hunting and trapping, bloody con-flicts were inevitable. Since the Beothuks did not possess firearms, they suffered heavy losses. Tuberculosis and other contagious diseases also reduced the Indian population. By 1811, the effects of warring and dis-eases had decimated the Beothuks, leaving them with fewer than 100 people. In 1829, Shawnadithit, the last Beothuk, died of tuberculosis.

Whaling by Europeans and Americans in Cumberland Sound, Hudson Bay, and Mackenzie Delta had similar impact on the health of the Inuit and dramatically reduced their population (Keenleyside 1990). Bone (1973) suggested that the low point in their population occurred around the turn of the twentieth century–several decades before the fur traders arrived. For example, the Inuit pre-contact population was estimated at 26,000; around 1900, it was 8,000; and by 1950, it was 10,000 (Bone 1973: 554). Contact with whalers played a role in two Inuit groups becoming extinct. The Sadlermiut inhabited Southampton Island in Hudson Bay. Crews of whaling ships would often spend the winter encamped there, employing the local Inuit in various tasks. This and other contact brought diseases. In 1902, a Scottish whaling ship visited Southampton Island and several Sadlermiut Eskimos fell ill, perhaps from typhus (Keenleyside 1990: 11). By 1903, every member of the tribe had died.

Mackenzie Inuit, numbering around 2000 in the nineteenth century, formed the densest Inuit population in Arctic Canada. A series of infec-tious diseases, beginning with scarlet fever and measles in 1865 and culminating with an epidemic of measles and influenza at the turn of the twentieth century, were major factors in the population decline of these Eskimos. By the 1930s, the Mackenzie Eskimos were no longer an identi-fiable group (Table 8.3).

Table 8.3
Mackenzie Eskimo Population Estimates, 1826 to 1930

Year	Population	Source
1826	2000	Franklin 1828: 68–228
1850	2500	Usher 1971a: 169–171
1865	2000	Petitot 1876a: x
1905	250	RCMP 1906: 129
1910	130	RCMP 1911: 151
1930	10	Jenness (1932) 1964: 14

SOURCE: Smith 1984: 349.

Recent Population Increase

Over the past fifty years, the population of Native peoples in northern Canada has increased dramatically. This increase began following World War II, with two events: the shift of Native people from the land to settlement (Vignette 8.2) and increased public expenditures on northern health care, allowing Native Canadians access to medical services. As a result epidemics ended; starvation was no longer a cause of death; and the probability of a mother and her new-born surviving greatly increased because most births now took place in hospitals or nursing stations. The immediate result was a sharp drop in the infant mortality and crude death rates. Since the 1950s, the natural rate of increase of Native peoples has been more than double the national average and much closer to the levels found in many Third World countries. These population increases among Native Canadians occurred across the nation. By 1986, the Native population may have reached its largest size ever–nearly 750,000, and just over 209,000 are found in northern Canada (Table 8.4).

The rapid increase in the Native natural growth rate has had a profound impact on the size of the northern population, although the enumeration of the Native peoples is in dispute. One reason is that not all Indian bands agreed to have their people enumerated. Census Canada estimates that some 45,000 Indians living on reserves did not participate in the 1986 census process. The impact of some 45,000 Indians has little effect on national or provincial figures but it can alter the population size of census divisions, particularly those with relative small populations. Manitoba census division 19 was probably just above 13,000 in 1986, and that population total shows a slight increase over the 1981 figure of 12,227.

In 1981, just over 413,000 Native people in Canada were recorded by the census. Five years later, the census recorded nearly 712,000 (though part of this increase is due to a change in the question). In the 1986 census, Canadians were asked: 'To which ethnic or cultural group(s) do you or did your ancestors belong?' A list of 15 ethnic groups was supplied, including Inuit, North American Indian, and Métis. Space was available for specifying other ethnic or cultural groups. Respondents were asked

Table 8.4
Population by Aboriginal Origins, 1986

Northern Census Divisions	No.	Per cent
Yukon	4,995	21.4
Northwest Territories:		
Baffin Region	8,120	81.5
Fort Smith Region	9,165	37.7
Inuvik Region	5,515	65.8
Keewatin Region	4,415	88.8
Kitikmeot Region	3,315	88.5
British Columbia:		
Kitimat – Stikine Regional District (49)	9,210	23.4
Bulkley – Nechako Regional District (51)	5,400	14.5
Fraser – Fort George Regional District (53)	6,525	7.3
Peace River – Liard Regional District (55)	5,890	10.3
Stikine Region (57)	415	20.5
Alberta:		
16	8,260	17.0
17	14,475	29.7
Saskatchewan:		
18	18,975	75.1
Manitoba:		
19	8,105	88.9
21	7,160	30.0
22	18,180	59.6
23	4,940	48.2
Ontario:		
Cochrane (56)	4,195	4.5
Algoma (57)	7,360	5.6
Thunder Bay (58)	9,430	6.1
Kenora (60)	10,865	20.7
Quebec:		
Abitibi (84)	3,420	3.7
Lac-Saint-Jean-Ouest (90)	2,280	3.7
Chicoutimi (94)	1,810	1.0
Saguenay (97)	6,120	5.9
Territoire-de-Nouveau-Québec (98)	14,930	40.2
Newfoundland:		
9	350	1.4
10	5,460	19.0
Totals	209,280	20.1

SOURCES: Statistics Canada, Special Print-Out.

Table 8.5
Aboriginal Population by Ethnic Groups, 1986

Ethnic Groups	Population	Per cent
North American Indian only	286,225	40.2
North American Indian and Non-Aboriginal Origins	239,400	33.6
Métis	59,745	8.4
Métis and Non-Aboriginal Origins	68,895	9.7
Inuit	27,290	3.8
Inuit and Non-Aboriginal Origins	6,175	0.9
Other Multiple Aboriginal Origins	23,995	3.4
Total	711,725*	100.0

*Census Canada estimated that another 45,000 Natives were not enumerated in the 1986 census, making a total of around 760,000.

SOURCE: Census Canada 1986, Aboriginal Peoples Output Program, Table 1.2.

to select as many origins as applied and therefore multiple responses were permitted. Since many combinations of responses by Aboriginal peoples occurred, it was difficult to tabulate these responses to obtain (a) the total population of Aboriginal peoples and (b) the totals for ten major groups such as Inuit. In 1986, 711,720 persons reported having Aboriginal origins, representing about 3% of the total population in Canada. Approximately half (53%) reported a single origin, i.e., Inuit, while the remainder (47%) said that they had a mix of both aboriginal and non-aboriginal origins.

Native leaders and some academics believe that the number of Native Canadians is much higher than the census figures. Wonders (1987: 664) reported that Asch (1984: 3) and Hamelin (1979: 213) both felt that the Native population around the late 1970s was probably over 800,000. Some five years later, McMillan (1988: 1) declared that 'over a million Canadians today can claim at least partial native ancestry'. Whatever the actual population of Natives, the 1986 census enumeration asked Canadians to state their ethnicity. This approach means that ethnicity is based on the respondent's perception of his or her ancestral background. The number of Aboriginal peoples by ethnic groups in 1986 is displayed in Table 8.5.

Population Distribution of Northern Aboriginal Peoples, 1986

In 1986, there were nearly 210,000 Native Canadians living in northern Canada. They were unevenly distributed across the Canadian North. The vast majority (nearly 90%) resided in the Subarctic. Table 8.7 shows the distribution of Aboriginal Canadians in three political units: the territories, the eastern provinces, and the western provinces. At the more detailed level of census divisions, the largest number of Native people

Table 8.6
Inuit Population by Province/Territory, 1961 to 1986

Province/Territory	1961	1971	1981	1986
Northwest Territories	7,977	11,400	15,495	17,385
Quebec	2,467	3,755	4,775	6,470
Newfoundland	815	1,055	1,365	1,810
Ontario	212	760	565	680
Alberta	85	135	315	300
Yukon and other provinces	279	450	685	650
Totals	11,835	17,550	23,200	27,295

Table 8.7
Aboriginal Population Distribution, 1986

Macro Census Regions	Population	Per cent Aboriginal	Per cent Regional
Territories	35,525	17.0	46.9
Eastern Provinces	66,220	31.6	6.9
Western Provinces	107,535	51.4	25.5
Total	209,280	100.0	30.1

are in Saskatchewan's census division 18 (18,975), Manitoba's census division 22 (18,180), Quebec's Territoire de Nouveau Quebec 98 (14,930), and Alberta's census division 17 (14,475). The northern census divisions with the highest percentage of Aboriginal peoples to the total population are Manitoba's census division 19 (89%), and the Northwest Territories' arctic census divisions, Keewatin Region (89%), Kitikmeot Region (89%), and Baffin Region (82%).

City Relocation

Over the past thirty years or so, a steady stream of Indian and Métis peoples have moved to large cities. Up to the 1986 census, Aboriginal peoples living in the Subarctic and Arctic are believed to have formed only a very small portion of this migration. Certainly, this statement is true for the Inuit whose migration to southern Canada was very small. Past census data describing the place of residence of Inuit provide a measure of the size of this migration. During this time, the percentage of Inuit living in southern Canada ranged from 4.8% in 1961 to 7.7% in 1971. In 1986, 6% of all Inuit lived outside their arctic homeland.

Settlement Life

After 1947, Native people of the Canadian North with the encouragement of governments began to settle around trading posts. Public monies were allocated to provide new services and to build a community infrastructure, including schools and nursing stations. Family allowances, old age pensions, and public housing were also available at these centres. In contrast to the 'grub stake' devised by the Hudson's Bay Company to keep trappers on the land, the federal government 'paid' them to stay in settlements. The rationale behind this was that (1) food supplies and medical care would be readily available in settlements, and (2) many northern natives could no longer support themselves on the land because of the diminishing numbers of wildlife, particularly caribou.

Vignette 8.2

Prior to 1940, the Cree people were scattered over a vast area of the boreal forest in northern Quebec. This 150,000–square-mile area provided food from hunting and cash from selling furs produced by hunting. People had lived in extended family hunting groups of about twenty people during the winter, but had joined together at the trading post for the summer. Some thirty years later, about one-third of the Cree of northern Quebec resided year-round in villages. By 1990, virtually all live in villages, though many continue to pursue a hunting/trapping economy.

SOURCE: Salisbury, 1986: 18.

Now, several generations later, young native adults have been born and raised in settlements. Many have yet to acquire sufficient education to obtain permanent jobs and participate in a wage economy; even if they did, there are far too few jobs available in Native settlements. Unlike communities in southern Canada or resource towns in the North, Native communities have no viable economic base and so offer few employment opportunities. Many of the social problem found in Native communities are exacerbated by high levels of unemployment.

Demographic Issues, Settlement Living, and Social Challenges

Since the 1950s, Native population in the North has increased rapidly. Coupled with this population explosion, relocation to settlements unleashed a series of social problems. Adjustment to some of the ensuing problems did occur but others proved more difficult. Sometimes these adjustments were beyond the capacity of the group and, as in the case of the Dene who were relocated to Churchill, social disintegration occurred (Dickman 1969). Later, the Dene returned to a bush settlement on North Knife Lake, Manitoba. The tragic experience at Churchill has been repeated at a number of other settlements, including Grassy Narrows, a

Vignette 8.3 Settlement of the Caribou Inuit

Before the late 1950s, the Caribou Inuit (Ahiamiut) lived inland and hunted caribou. When their chief source of food, the barren-ground caribou, became scarce, they fell into difficult times. Their remote location and their independent life-style made communications with the outside world irregular, and therefore little help was available. In the 1950s, their main contact was through the Department of Transport weather station at Ennadai Lake. When word began to reach the outside world that the Ahiamiut were starving and that many had died, the federal government attempted to rescue the survivors by resettling them in coastal villages such as Eskimo Point. Williamson (1974: 90) reported that starvation had reduced the population of Caribou Eskimos from about 120 in 1950 to about 60 in 1959.

SOURCE: Williamson 1974: 90.

small Ojibway settlement in northwestern Ontario. The unravelling of Native social fabric does not always occur in settlements, but almost all are troubled by social problems. At Grassy Narrows, Shkilnyk (1985) describes the massive and extraordinarily rapid change in these people's way of life caused by a combination of relocation and mercury pollution. These two events led to the unravelling of the tribe's social fabric and their personal morale–dramatically reflected in the high number of violent deaths, suicides, and alcoholism. While similar social problems can be found in other Native communities, these two communities highlight the appalling social conditions brought on by the relocation of Native people from a bush environment to an urban one. In the case of Grassy Narrows, the stress brought about by relocation was added to by the presence of high levels of mercury in fish which is the chief source of food for these Indians.

In spite of the stressful nature of relocating into settlements, the size of these communities is increasing and local government has arisen. As these settlements continue to grow, a new set of problems emerge, such as a source of uncontaminated drinking water and the demand for more public housing. Native settlement councils are grappling with these and other questions, and the emergence of local government may be a sign that Natives are not only playing a role in determining the nature and character of settlement life but are also taking charge of their destiny. Salisbury and Stenbaek have both identified a resurgence that may be related to the growth of Native self-government. In the case of Cree in northern Quebec, Salisbury believes that a sense of regional identity has emerged. This Cree regionalism is strengthened by local control of band affairs and the newly formed inter-community links associated with the James Bay settlement. Further north, Stenbaek believes that the Inuit are coping with the social pressures exerted by non-Natives and she sees a rejuvenation of their culture.

Natural Rate of Increase

The natural rate of increase of Canada's northern Aboriginal population is well above the national average. Modern health care keeps the death rate low and the birth rate for Native Canadians remains high. Crude death rates in the North are under ten per thousand persons and crude birth rates are around thirty per thousand persons.

Over the last thirty years, the natural rate of increase of Native people has declined but it is still much higher than the national average. In 1985, for example, the Aboriginal peoples of Canada had a rate of natural increase of 2.8%, nearly four times greater than the national figure of 0.8% (Hagey, Larocque, and McBride, Part I, 1989: 22). For the same year, the Inuit growth rate was 2.5%.

This high rate of increase has two implications. The first deals with the percentage of Natives in the northern population. Given the expected continuation of high rates of natural increase for Native people and assuming no migration of non-Natives, the percentage of Natives in the northern population is expected to increase from 20% in 1986 to around 25% by the turn of the century. Within the North, regional differences may occur. The greatest increase in the proportion of Native people to the total population is likely to occur in the Subarctic where resource development is declining, causing a rise in out-migration of non-Natives. In contrast, any drop in the percentage of Native people is likely to occur in regions where resource development is expanding and there is an in-migration of non-Natives.

The second implication is an increase in the proportion of Aboriginal peoples in the working-age group (15 to 64 years). From 1981 to 2001, the age-structure of the Inuit population is expected to change, resulting in a 7% increase in the size of the Inuit potential labour force by 2001. Similar increases are expected by 2001 for Indians and Métis living in the Subarctic. With an already high unemployment rate among Native adults and a declining fur economy, the northern Native peoples but particularly those living in the Arctic will face an even more difficult future.

Freeman, in 1971, expressed concern about the imbalance between Inuit population size and their resource base, writing that 'a rapid rate of population growth generally prevents any successful attempt to remedy the prevailing unfortunate economic situation'. Since then, the population has continued to increase and the Arctic economy remains stagnant. With little prospect of regional or community economic growth, the Arctic (and other areas in the North) is troubled by the so-called Malthusian 'population trap'. This trap occurs when a country or region has a higher rate of population increase than the rate of food production. In classical Malthusian terms this trap eventually forces people to live at subsistence levels of income; in the Canadian North, it would mean greater reliance on store food and social assistance. This population problem, explored by Irvin (1988) and Freeman (1971), is double-edged–an extremely high rate of population increase and an extremely low creation of jobs in the Arctic. Given the predicted population increase of the Inuit population in the Northwest Territories from 25,000 in 1981 to an estimated 40,000 in 2001 (Hagey, Laroque, and McBride 1989: Table B), the current economic and social problems facing Inuit communities are likely to intensify. During this same time, the Inuit potential labour force is predicted to increase substantially as the population ages. Federal government officials estimate an increase in the Inuit potential labour force from 54% in 1981 to 62% in 2001 (Hagey, Laroque, and McBride 1989: 9, 15).

Disadvantaged People

The descendants of the first inhabitants of the North are in a state of transition, moving from a land-based to a settlement economy. In this transition, they face many challenges. The outcome of this process is far from clear. One possibility is that a culture of poverty will emerge (Frideres 193:185); another is that a vibrant Native culture will prevail (Salisbury 1986). Most socio-economic statistics paint a very dismal picture, supporting the culture of poverty model. But this picture is like looking in a rear-view mirror–it shows the past but provides little indication of the future. Whatever the future, it depends not upon governments but on Native northerners seizing economic opportunities, re-establishing social cohesiveness, and striving to succeed in their endeavours. One barrier to be overcome lies in their location–many live in isolated communities with few job opportunities and little incentive to stay in school. Such an environment lends itself to a dependency on social welfare. One possible solution is to create a land-base option, one providing financial incentives to stay on the land.

Canadians often forget that Native Canadians have a special relationship with the federal government and that relationship is defined by the Indian Act. Rather than seeing themselves as 'disadvantaged', Native Canadians can rightfully consider themselves 'Canadians plus'.

Treaty Indians, for example, are legally different from other Canadians. They may live on reserves where they are governed by a band council. In the provincial North, such reserves are surrounded by Crown land and Treaty Indians often hunt and trap on these lands. In the territories there are few reserves, but Indians, Inuit, and Métis hunt and trap on Crown land. The general pattern of life consists of two parts–hunting and trapping on surrounding lands and residing on a reserve where their basic needs such as housing are met by various federal programs. While native settlements and reserves tend to isolate Natives from the rest of Canadian society, they offer a place where Indian values, languages, and customs are still practised.

Given this lifestyle and special relationship, the notion of 'ethnically blocked mobility' arises; that is, Native Canadians may value educational and occupational achievements less than other Canadians because of cultural differences such as the sharing ethic and because of perceived or experienced discrimination (Lautard and Guppy 1990: 191).

The level of Native development was measured by Stewart (1987) using the 1981 Native Summary Tape to compile a series of development indicators at the census division level. Each of the 140 census divisions was classified into four development levels–extremely high, above average, below average, and extremely poor. The 29 northern census divisions did not rate well–nearly 80% fell into the below average and extremely poor categories. Eight of the 24 census divisions classified as having extremely poor Native development are located in the Subarctic and Arctic, namely Baffin 04, Saskatchewan 18, Manitoba 19, 22, and 23 and Quebec 84, 97, and 98.

Past efforts by governments to address the economic problems facing northern Native Canadians, whether it be job training, public housing, or health care, have not altered the fact that Native Canadians continue to be the most disadvantaged group in Canada. As a group, Aboriginal peoples remain more likely to be unemployed, to live in housing in need of major repairs, and to die at an earlier age than other Canadians. Individual federal, provincial, and territorial programs have improved their living conditions but have not solved the underlying problems; they may well have increased Native dependency on government. One such program is housing. The federal and provincial governments have invested millions of dollars into home ownership programs and rental programs for Native people. On the one hand, subsidized housing is needed, for many Natives would otherwise be living in substandard housing, a mixture of self-built shacks and log cabins. Such shanty-town type housing existed in the 1950s and 1960s and the ensuing social and health problems were unacceptable; hence a series of public housing programs was initiated. On the other hand, these programs have created new problems which require more public intervention. For example, the vast majority of Native families who are involved in self-ownership housing programs cannot afford to maintain their houses and these houses soon become run down (Bone and Green 1983). Government is then faced with repairing the units or with providing replacement housing.

Educational Attainment

Today, a measure of Native disadvantage is low educational attainment. Since many Native northerners over the age of 40 have little or no schooling, their educational attainment (as defined by the Canadian educational system) is extremely low. These adults would be classified as functionally illiterate, defined as having less than Grade 9 education.

Table 8.8
Educational Attainment in the Northwest Territories, 1989 (%)

Education (completed)	Inuit	Inuvialuit	Dene	Métis	Non-Natives
<Grade 9	62	48	55	25	4
Grades 9–11	18	20	18	29	13
Grade 12	3	7	7	18	27
University Degree	—	1	1	3	23
Certificate	14	23	19	25	33
No Response	3	1	1	—	1
Total	100	100	101	100	101

SOURCE: Government of the Northwest Territories, Bureau of Statistics, 1990: 26 and 28.

Within traditional Native pursuits such as hunting and trapping Canadian education is not a significant factor. However, within the wage economy, most employers require that prospective employees have graduated from high school. There are at least three categories of an adult person's employability: (1) those with less than grade 9 education are the least employable; (2) those with some high school education are restricted to manual jobs; (3) those who have graduated from high school have the greatest job opportunities. With the possible exception of manual employment, young adults who are seeking their first job and who have not graduated from high school have effectively removed themselves from much of the job market and certainly from the opportunity to gain permanent employment.

Low educational attainment is one reason why so few Native Canadians are employed. For example, the national rate of functional illiteracy was 17% in 1986 while the rate for Aboriginal peoples was more than double that figure (Hagey, Larocque, and McBride 1981: 27–28). In 1989, education levels by ethnic groups emphasizes this national pattern (Table 8.8). The percentage of high school and university graduates ranges from a high of 50% for non-Natives to a low of 3% for Inuit. The percentage for the other Native groups are 21% (Métis), 8% (Dene), and 8% (Inuvialuit).

Among the Aboriginal peoples, the Inuit have one of the highest rates of fuctional illiteracy, at 53%. Similarly, the proportion of Canadians of Native ancestry who have graduated from high school was well below the national average of 56% in 1986. Such levels of schooling effectively prevent many Natives from entering the wage economy, where many employers require at least a Grade 12 education. Both territorial governments continue to hire applicants from southern Canada because they have the educational qualification for the jobs, whether it be an accountant, school teacher, or an electrician.

Table 8.9
Illiteracy Rates (per cent) for Canadians, Status Indians, and Inuit, 1981 and 1986

Group	Minimum of Grade 8		Minimum of Grade 12	
	1981	1986	1981	1986
Canada	20.1	17.3	52.1	55.6
Status Indians	39.1	37.2	26.4	27.6
Inuit	61.3	53.0	18.9	22.2
Aboriginals	37.3	25.9	28.3	39.7

SOURCE: Hagey, Larocque, and McBride, 1989. *Economic Conditions*: 27–8.

There are several reasons why there are so few Native high school graduates. One explanation is an historic one. Public schooling was first accessible to many Native children some forty years ago. In the Arctic, smaller communities did not obtain a federal school until the 1960s (Clyde River, Igloolik and Pond Inlet in 1960; Grise Fiord and Padloping in 1962, Lake Harbour in 1963, and Hall Beach in 1967). Most northern schools do not offer Grade 12 and only larger communities have high schools. For example, in 1989, there were 76 schools in the Northwest Territories but only eight offered full senior high school programs (two of which are located in Yellowknife). The high schools are located in seven regional centres (Fort Smith, Hay River, Inuvik, Iqaluit, Pangnitung, Rankin Inlet, and Yellowknife). A second reason is linguistic. Initially, the curriculum was entirely in English and was patterned on material used in southern Canadian schools. Grade 1 started with a 'Dick and Jane' reader and the subject material was often meaningless to Inuit and Indian children. Many of those entering Grade 1 were exposed to English for the first time, compounding the difficulty of their educational experience. While the northern schooling system has incorporated Native languages and history into its curriculum, these languages occupy secondary position. Dorais (1989) argues that the relegation of Inuktitut to a secondary position has resulted in many younger Inuk being unable to speak their mother tongue. This loss, called diglossia, occurred among the Dene some twenty years earlier; other Indians have suffered the same language loss prior to the 1950s. Such losses affect self-identity and self-esteem. Another explanation is that the curriculum reflected western cultural values and economic norms. These values and norms stressed individual achievement, competitiveness, and a reward system. Since such a cultural orientation often conflicted with values learned in the home, Native youngers were living in two cultural worlds: their parents' version and that presented at school.

Adjusting to an education system set in a different culture is a formidable task. Other factors affecting performance include crowded housing, the high teacher turnover, and the limited number of Native role models. However, the lack of jobs in local communities is the most important factor because it means that there is little chance of obtaining employment even with a high school degree.

Health

Health is another indicator of the well-being of a society. It is affected by economic, social, and demographic factors. Within Canadian society, the health level of Aboriginal peoples is lower than that found in Canadian society in general. The best single measure of health is life expectancy. Life expectancy for Status Indians is approximately ten years lower than the national average. In 1981, Indian males had a life expectancy of 62.4 years compared to 71.9 years for Canadian males. Simlarly, Indian females had a life expectancy of 68.9 years, while the national figure for all females was 79 years. Infant mortality rates provide another indicator of the health of a society. Again the national rate for Natives is well above the national average of eight infant deaths per thousand births. Within the Northwest Territories, the highest infant mortality rates are found in the Inuit population, though this rate has declined from 38 per thousand births in 1981 to 28 per thousand births in 1986. Life expectancy and infant mortality, while measured in terms of demographics, is shaped by a composite of socio-economic factors such as health care, working conditions, and quality of housing, and psychological factors, such as self-esteem, pride, and purpose in life.

While there are signs that Native health level is improving, it remains well below that of other Canadians. Why is it that Indians, Inuit, and Métis do not enjoy the same level of health as other Canadians? Accessibility to medical services is one of the factors. Most Canadians live in urban settings with a wide variety of health care institutions and workers close at hand. In the remote communities, access to the full range of health care is not as readily available. For example, there are only four hospitals in the Northwest Territories and only one of these four is found in the Arctic. Low incomes, diet, and alcohol/drug abuse may also affect the health of Canadians of native ancestry living in the North.

Summary

In northern Canada, significant demographic and social changes have taken place in the post-World War II period. For Aboriginal Canadians, these changes were triggered by a change in government policy which encouraged them to relocate to settlements. Living in settlements allowed Native Canadians greater access to a number of public services, including health care, and to an urban way of life. During this time, the natural rate of increase of Native Canadians has been much higher than the national average. The highest rates are found among the Inuit people and these are close to three times greater than the national average.

Figure 8.1
Inuit Demographic and Social Indicators, Northwest Territories

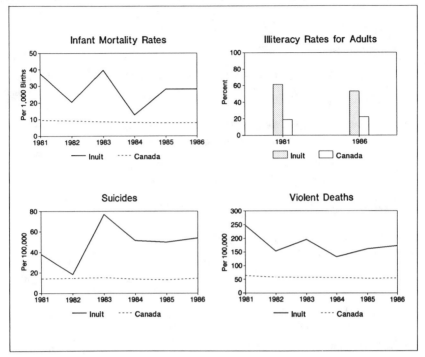

SOURCE: Hagey, Larocque, and McBride 1989.

Settlement living has also brought about social changes, some of which have had an adverse impact on Native culture, particularly the replacement of Native languages with English. These social changes can be viewed as part of a 'modernization' process of the North and its Native people. Education is one such an example but the results so far have not served young Native adults well because their education levels are well below those found in the larger Canadian society. Settlement living and the resultant social problems found in these centres has left Native people in a disadvantageous position relative to the rest of Canadian society. There is no doubt that the development of northern resources and the accompanying 'top-down' approach reinforces Native dependency, isolates them from their culture, and limits Native control over their community. In a word, Native peoples have been 'marginalized' by the processes of economic development (few jobs in Native communities and insufficient cash income from trapping) and social modernization (extreme pressure to conform to the dominant society's values and to participate in its economic system). Native people will continue to be a disadvantaged people until they can gain more control over their lives and be responsible for their decisions. Even then, the limited economic potential of Native communities may confine residents to lives of underemployment and low incomes.

Suggested Readings

Bone, Robert M. and Milford B. Green, 1983. 'Housing Assistance and Maintenance for the Métis in Northern Saskatchewan', *Canadian Public Policy* 9 (4): 476–86.

Damas, David, ed., 1984. 'Arctic'. *Handbook of North American Indians*, William C. Sturtevant (general editor). Washington: Smithsonian Institution.

Helm, June, ed., 1981. 'Subarctic'. *Handbook of North American Indians*, William C. Sturtevant (general editor). Washington: Smithsonian Institution.

McMillan, Alan D., 1988. *Native Peoples and Cultures of Canada: An Anthropological Overview*. Vancouver: Douglas & McIntyre.

Robitaille, Norbert and Robert Choinière, 1987. 'The Inuit Population of Canada: Present Situation, Future Trends', *Acta Borealia* 4 (1–2):25–36.

Salisbury, Richard F., 1986. *A Homeland for the Cree: Regional Development James Bay, 1971–1981*. Kingston: McGill-Queen's University Press.

Stenbaek, Marianne, 1987. 'Forty Years of Cultural Change Among the Inuit in Alaska, Canada and Greenland: Some Reflections', *Arctic* 40 (4): 300–9.

Wonders, William C., 1987. 'The Changing Role and Significance of Native Peoples in Canada's Northwest Territories', *Polar Record* 23 (147):661–71.

Chapter 9

NATIVE ECONOMY AND LAND CLAIMS

The Canadian North takes the form of a resource hinterland where economic decisions are resolved by supply and demand. Within this hinterland economy, another economic system, the Native economy, exists. It serves another way of life, one that was a land-based economy but now blends settlement and land-harvesting activities. The present Native economy has evolved from fur-trade days, but its roots reach back much further in time to a subsistence hunting economy, and many of its attitudes and actions are governed by values formed in earlier times. It is these Native cultural values that maintain the economic system and allow it to modernize within these cultural constraints. In this sense, the comprehensive Native land claims can be interpreted as means to ensure that the Native economy survives in a new, more viable form, in which Native peoples have sufficient land, capital, and political control over their regional economy.

One aspect of the Native economy is its strong ties to the land; another is the growing need for cash income. These two elements of the Native economy represent its strength and its weakness. Land-based activities have not been able to generate enough cash income to support settlement life style. This life style requires the purchase of goods and services common to all settlement dwellers in the Canadian North. It also requires the 'modernization' of hunting and trapping activities. Native peoples have turned to wage employment for most of their cash income, with lesser amounts accruing from transfer payments and fur sales. In 1985, for example, employment income accounted for 74% and 80% of the total income received by Aboriginal peoples in Yukon and Northwest Territories respectively (Table 9.1). Total income does not include the 'cash' value of country food which may, on average, form as much as half of all the food consumed by Native northerners.

Before the 1950s, the Native economy was driven by the fur trade. Furs were bartered for store goods and equipment, with the objective of making living on the land easier for the extended-family hunting unit. Items of trade would include hunting and trapping weapons and equipment; domestic goods such as blankets, tents, and cooking and sewing utensils; basic foodstuffs such as flour, tea, and sugar; and such non-basic items as tobacco. Whether in the bush or on the tundra, the Native economy was concerned with harvesting and consuming food from the land. In this subsistence economy, material wealth was limited to those weapons, tools, and other belongings that could easily be carried to the next hunting, fishing, or food-gathering location. The basic social unit was the extended family, and cultural values such as co-operation and sharing were instrumental in group survival. But beyond these values, the land and animals took on a spiritual significance and this Native version of animism helps explain Native peoples' commitment to the land.

Once Native people moved to settlements, their lives began to change as they became enmeshed in an urban life style. Native social and economic order remains in a transitional state. The process of adjustment to these new circumstances is reshaping relations with non-Native society. A new Native economy is emerging, and much of this activity is taking place within the structure of Native development corporations. But most of all, settlement living has intensified social and economic interaction of Native people with non-Native controlled institutions, and increased pressure to conform to non-Native norms. One such norm is the minimum level of education for many job applications, a high school diploma. Settlement living also created a demand for non-Native material possessions such as a house, truck, washing machine, electric stove, store clothes, and food. Social arrangements were altered as families no longer lived on the land for most of the year; now men, not families, form trapping/hunting groups and spend relatively short periods of time on the land. The economics of trapping changed too, as settlement living required modern transportation such as snowmobiles, trucks, and even chartered aircraft rather than dog teams. While the cost of trapping has increased in the 1980s, the price of fur has not kept pace. For the Inuit hunter, the collapse of the seal skin market has removed an important source of cash income; the possibility of a similar drop in the demand for furs exists. For example, the 1988–89 value of fur in the NWT was nearly 30% less than the previous year (Table 9.7).

Native trappers have faced similar dire circumstances before and yet

Table 9.1
Aggregated Income for Aboriginal Peoples 15 Years and Over, 1980 and 1985

Income Source	Yukon		Northwest Territories	
	1980	1985	1980	1985
Employment	82.3%	74.2%	79.5%	79.8%
Self Employed*	3.6%	4.8%	5.3%	3.0%
Public Transfers	12.5%	19.0%	14.2%	16.0%
Other	1.6%	2.0%	1.0%	1.2%
Total	100.0%	100.0%	100.0%	100.0%
Aggregated Income ($millions)	18.9	33.0	91.3	172.8
Per capita Income	$4,674	$6,605	$3,453	$5,661

*mainly income from trapping
SOURCE: Larocque and Gauvin 1990: 95.

trapping has persisted. Only a decade ago, Dacks (1981: 13) observed that 'while only a minority of native people support themselves by hunting, fishing, and trapping, these activities remain very important to native people.' A strong attraction to the land persists. This attraction is based on Native spiritual links to the land and to practical considerations, namely that the land is a source of highly valued wild foods and fur-bearing animals.

Wage Employment

Statistics reveal that Native participation in the wage economy is increasing across Canada and in the North. As well, Native income from employment earnings is rising. The per capita incomes for Native adults in Yukon and Northwest Territories rose from 1980 to 1985 (Table 9.1). Even so, in each territory the non-Native per capita income is substantially higher (2.2 times in Yukon and 3.4 in the Northwest Territories). This income gap indicates that Native workers still occupy lower paying jobs, although Natives employed in regional centres have much higher incomes than those in smaller places where few full-time jobs are available. In 1980, for instance, the average household income of Inuit in regional Arctic centres of the Northwest Territories was nearly $25,000 while households in smaller settlements received only $17,300 (Usher and Wenzel 1989: 18). Yet surveys of Native northerners indicate a strong interest in the wage economy. For instance, the 1989 GNWT labour force survey (1989: 12) revealed that nearly 17% of those Native adults interviewed were actively seeking jobs and another 39% were employed. The report contained another revealing statistic, namely that just over 63% had worked during 1988.

Scholars interpret such statistics differently. Some see the growing involvement in the wage economy as part of the modernization of Native

Table 9.2
Aboriginal and Non-Aboriginal Per Capita Income, 1980 and 1985

	Yukon		Northwest Territories	
Income Source	Aboriginal	Non-Aboriginal	Aboriginal	Non-Aboriginal
Employment	$4,901	$12,175	$4,519	$17,247
Self-Employment	318	793	167	702
Government Transfers	1,253	1,021	980	631
Other Income	132	805	68	623
Total	6,605	14,795	5,661	19,202

SOURCE: Larocque and Gauvin 1990: 95.

society, while others see it as a means to support land-based activities. Berger (1977: 93–100) argued that traditional activities cannot be understood from a commercial perspective because Natives engage in these activities as a way of life, stressing different cultural values than those found in the rest of Canadian society. Dacks (1981: 13) supports this line of reasoning by observing that 'many native people who are employed– and therefore do not have an economic need to do so–continue to hunt, fish, or trap in their spare time or in interludes between periods of wage employment.' In western economic terminology, people may choose a less attractive economic option because it results in greater personal satisfaction. Thus, a sub-optimal rather than an optimum economic decision may occur. The 'cost' of that decision is the difference between the higher economic gain that could have been derived from the optimum economic decision and the actual monetary gain resulting from the sub-optimal decision. This difference is called 'psychic income'. A Native person, for instance, may prefer a mix of summer employment and winter trapping/hunting even though more money could be secured by full-time employment.

In analyzing the 1984 Northwest Territory Labour Force Survey, Stabler (1989a: 829) concludes: (1) that Natives now hold over 6,000 jobs of which 38% are managerial and skilled jobs; and (2) that the increase in the number of employed Native northerners over the last 30 years is significant. In another article, Stabler (1989b: 295) states that the same 1984 survey indicated that 42% of those holding jobs also engaged in traditional activities. He suggests three hypotheses to explain the continuing participation in hunting and trapping activities. The traditionalist view is that Native people prefer hunting, trapping, and fishing activities over wage employment. The modernist position is that wage employment is preferred and that Natives engage in hunting, trapping, and fishing activities as a 'last resort'. The third perspective is held by the culturalist who believes that Natives engage in hunting, trapping, and fishing activities in order to maintain their cultural heritage. The survey data do not

Table 9.3
Social Assistance Payments, Northwest Territories 1982 to 1989 in $ millions

Year	Amount	Year	Amount
1982	$8.2	1986	$12.1
1983	9.4	1987	16.9
1984	9.0	1988	19.4
1985	10.8	1989	20.5

SOURCE: Government of the Northwest Territories, Bureau of Statistics 1990: 23.

unequivocally support any of the hypotheses, perhaps because the data represent individuals rather than extended family units which would allow a combination of economic activities. Such a combination within a family unit would be an effective adaptation, allowing both an increase in cash income and a continuation of traditional pursuits. These traditional pursuits produce country food which, by reducing the need for store food, could be seen as an economic gain. Other 'economic' reasons why Native northerners continue to hunt and fish are the high cost of store food, and a preference for fresh fish and game over frozen meat, fish, and poultry available in the local store.

In spite of a growing involvement in the wage economy, the average Native income remains low compared to non-Native income. In 1985, the per capita income for an Aboriginal Canadian living in the Northwest Territories was just under $5,700. Wage employment accounted for nearly 80% of this income while public transfers made up most of the remainder (Table 9.1). From 1980 to 1985, average income for Native employees in the two territories increased significantly (from $3,454 to $5,661 in the NWT). Even so, these incomes remain far behind those earned by non-Native persons. In the case of the Northwest Territories, the 1985 per capita income for non-Natives was 3.8 times greater than Native per capita incomes; in Yukon, the difference was 2.5 (Larocque and Gauvin 1990: 95). Over this five-year period, the income gap between the two groups has actually increased. Given this substantial and apparently growing income difference, it is surprising that per capita government transfer payments to each group vary so little. For example, in the Northwest Territories, government transfer payments to non-Natives was $631 while to Natives it was $908, a difference of only 44%; in Yukon, the difference was only 23% (Larocque and Gauvin 1990: 95). Yet, in percentage terms, transfer payments form a large share of personal income of Native people, suggesting an economic dependency on monies provided by family allowance, unemployment insurance benefits, and welfare payments. In 1985, transfer payments comprised nearly one-third of the income obtained by Aboriginal peoples in Canada; the figure for Inuit living in the Northwest Territories was slightly lower at just over one-quarter (Hagey, Larocque, and McBride, II: 33). In the Northwest Territories, social assistance

Table 9.4
Per cent of Population 15 Years and Over in the Labour Force, 1981 and 1986

Group	1981 (%)	1986 (%)
Registered Indians on Reserves, Canada	39.4	43.3
Registered Female Indians on Reserves, Canada	27.8	32.3
Native, Yukon	44.6	44.8
Native, Northwest Territories	39.6	40.2

SOURCE: Larocque and Gauvin 1989:75, 79 and 93.

payments have increased in the 1980s from $8.2 million in 1982 to just over $20 million in 1989 (Table 9.3). Two reasons for this sharp increase are (1) the growing number of young Native people with no visible means of support; and (2) the decline of the trapping industry.

Government income figures for northern Native individuals and families fail to account for two other sources of their economic support. One is country food. Since many Aboriginal northerners obtain much of their food supply from the land, they are able to reduce their purchase of store foods. Another comes in the form of federally funded goods and services available only to Aboriginal Canadians, among them housing, dental services, and post-secondary education.

During the 1980s, the number of Native employees increased but relatively few held permanent jobs (Tables 9.4 and 9.5). Seasonal employment is highest during the summer when construction projects take place. The findings of the 1989 NWT Labor Force Survey illustrate this employment pattern. In 1988, some 12,430 Native persons were employed by firms and public agencies but only 31% worked for more than 26 weeks (Table 9.5). The seasonal character of the 1989 employment shows a summer high and a winter low: the percentage of Native persons employed reached a low of 35% in January (6,829 workers) and a high of 46% in August, with 9,055 (Government of the Northwest Territories, Bureau of Statistics 1990: 7–8). Within the Northwest Territories, the percentage of the Native potential labour force employed varied, with the Inuvik Region having the highest percentage at 67% while the Kitikmeot Region has the lowest percentage at 60%. The higher Native employment figure for Inuvik Region may be attributable to the Inuvialuit Land Claim settlement.

Wage Employment and Air Commuting

Few wage-employment opportunities exist in Native communities. One solution is to move 'surplus' population to areas that need workers. Another is to stimulate local development. The migration model is theoretically sound, but assumes that the migrating labour force has education qualification and job skills demanded by employers in the region seeking

Table 9.5
Persons who Worked during 1988 by Region and Ethnic Group, Northwest Territories

Region	Persons 15 Years & Over	Worked in 1988 Number	%	% Worked More Than 26 Weeks
Baffin Region	6,308	4,376	69	41
Native	4,898	3,108	63	30
Non-Native	1,410	1,268	90	80
Keewatin Region	3,195	2,090	65	35
Native	2,723	1,649	61	27
Non-Native	472	440	93	80
Kitikmeot Region	2,457	1,595	65	33
Native	2,136	1,288	60	25
Non-Native	322	307	95	84
Inuvik Region	5,531	4,167	75	49
Native	3,672	2,465	67	33
Non-Native	1,859	1,702	92	81
Fort Smith Region	17,159	13,987	82	62
Native	6,183	3,921	63	36
Non-Native	10,976	10,067	92	77
Total	43,650	26,215	76	51
Native	19,611	12,430	63	31
Non-Native	15,039	13,785	92	78

SOURCE: Government of the Northwest Territories, Bureau of Statistics, 1990: 23 and 26.

workers. In the North, most Native people live in small, isolated settlements. The migration solution would relocate Native peoples in regional centres and resource towns. Already, a small number of Native families and individuals have moved to larger, regional centres, but few have moved to resource towns. Government efforts in the 1960s to relocate Native families to resource towns have not worked well (Stevenson 1968, and Williamson and Foster 1975). Since the relocation of Newfoundlanders from isolated outports to regional centres, the notion of massive relocation is socially and politically unacceptable. Yet the basic problem of a misfit between the location of resource sites and Native settlements remains. Usher, speaking at a research seminar on the North associated with the Royal Commission on the Economic Union and Development Prospects for Canada, argued that: 'Looking for either mega-project (or even mini-project, such as mining) development as the solution for native employment is barking up the wrong tree. There is absolutely no evidence that native people are longing to move out of their communities to a few points where economic opportunity is so much greater' (Whittington 1985: 17).

An alternative approach to relocation to resource towns is air commuting, which offers Native workers access to the work place but allows their families to remain in their settlement. Recruitment of Native workers by mining companies has met with some success. In 1987, there were 385 Native employees working at fly-in mines; and during the construction of the Norman Wells Project, there were 538 Native project workers using an air-commuting system to reach the construction site. Native air commuters form a small but well-paid sector of the mining work force. In northern Saskatchewan, there were 320 Native air commuters in 1987, forming nearly 28% of the total northern mining workforce (Shrimpton and Storey 1989: 17). Figures for the Northwest Territories indicate that Native commuting employees comprise just under 10% of the air-commuting workforce. These figures represent a significant increase over past involvement of Native workers in the mining industry (Bone and Green 1984:12). Salaries are high and starting wages at Saskatchewan's uranium mines is around $35,000 a year.

While the economic rewards are substantial, the social impacts on Native rotational workers and their families are mixed. During the early 1970s Tough (1972: 85) wrote that 'if there is a uniquely northern social problem associated with mining, it is the failure to involve the indigenous people of the region in the economic activity. Too often they have been spectators only, or have been socially disrupted by the wave of development.' Some writers (Beveridge 1979 and Hobart 1989) argue that many of the social problems associated with moving to a company town are eliminated by allowing Native workers to live in their home community. Yet, there is a small but possibly growing number of Native workers who have relocated with their families to southern cities. While the information on this topic is sketchy, there are several reasons why such a move might take place. These reasons include 'push factors' such as social pressure on family while a parent is at the work site; economic pressure resulting from the high-income family surrounded by Native families with much lower incomes; and local pressure from the band office or settlement council regarding public housing and other items controlled locally. 'Pull factors' include minimizing travel time to the worksite by relocating to a major centre which serves as the principal origin/destination for the aircraft; and relocating to a city to take advantage of financial services, retail shopping, medical institutions, and educational facilities.

The record of northern and Native employment varies by region and by company. The Saskatchewan North has the best employment record with one company easily recruiting half its workforce from the North, most of them workers of Native ancestry. The three mines in the Northwest Territories have fared less well–only 14% of the workers at Nanisivik, Polaris, and Lupin mines reside in the NWT. Within this context, it is interesting to note that over half of the Norman Wells Project rotational workers employed by Esso and its subcontractors resided in the Macken-zie Basin of the Northwest Territories. The necessity of a northern air-

Table 9.6
Native Employment at Selected Mines, 1987

Mine	Native Employees	Per cent
Cluff Lake, Sask.	110	41.8
Rabbit Lake, Sask.	95	25.5
Key Lake, Sask.	110	22.6
Star Lake, Sask.	5	9.4
Polaris, NWT	26	9.3
Lupin, NWT	35	7.9
Detour Lake, Ont.	4	2.7
Total	385	19.4

SOURCE: Shrimpton and Storey 1989: 36.

commuting system to complement the main north-south one is a fact of doing business in the northern hinterland. The Saskatchewan Government, for instance, insists that the uranium mines employ residents from its small northern population, while the Territorial Government of the Northwest Territories places a similar responsibility on its mining industry.

Community Development

Another approach is to stimulate economic activities in Native communities. Currently, private investors see few opportunities in such places. The alternative to privately funded economic ventures is community development projects. Such projects normally involve local entrepreneurship, community support, and public financing. The attraction of community development is that economic control is vested with the community, wages and profits remain in the community, and local expertise will be developed (Robinson and Ghostkeeper 1987 and 1988). Community development, therefore, is a locally induced economic initiative, occurring within the context of a market system. The key factor is that local factors play a principal role. In this sense, the term 'locally based' describes the development's nature and purpose. Coffey and Polese (1984) described local development as a particular form of regional development but one that operates at a micro-spatial level and in which local factors play a central position. They proposed a four-stage model of local development: (1) the emergence of local entrepreneurship, (2) the 'take-off' of local enterprises, (3) the expansion of these enterprises beyond the local region, and (4) the achievement of a regional economic structure that is based upon local initiatives and locally created comparative advantages. In theory, community development has much appeal, but in reality such development is troubled by a small market and high cost of operation, resulting in high prices for goods and services. Local consumers' loyalty

---

Vignette 9.1 Formal / Informal Economies

Economic activities can be divided into two groups, the formal and the informal economy. The principal difference between the two economies is that those involved in the formal economy record their financial transactions and pay taxes to the different levels of governments. The activities of the 'hidden' or informal economy are not reported to government and taxes are not paid. The informal economy consists of two major components, one of which involves the exchange of money and the other which involves not money but bartering, sharing, or helping. The monetized informal economy is divided into legal activities and illegal ones (the underground economy). Informal businesses, called micro-enterprises, are small-scale and labour-intensive. Largely cash-based, these businesses are often operated out of a house or garage by people with little capital and a few skills. They occupy a market niche in a community. They represent positive adaptations by people who, because of limited education or skills, are unable or who do not want to seek regular employment. Wolfe argues that micro-entrepreneurs are motivated by both economic and social factors.

SOURCE: Wolfe and Convery, 1989.

---

to buying locally is often severely tested, causing some to purchase goods and services outside of their community. Given this situation, most forms of community development require annual operating grants to offset losses.

In many northern communities local Native residents or Native corporations operate small businesses. The growth of Native-run businesses demonstrates a new dimension to the 'traditional Native economy'. Those operated by a Native organization represent an attempt to create a Native business which will further the economic well-being of its members within a Native setting. These organizations range from construction firms to store management operations. Often the capital necessary to begin operations is obtained from the federal government. The smaller firms might better be described as examples of the 'informal Native economy'. For instance, the Chief of an isolated settlement without a store has purchased a stock of canned goods and stores them in his house. Band members can purchase these goods rather than obtaining them from a store located in a distant community. The Chief has purchased these foodstuffs as a service to his tribe rather than as a means of making money. A Native women operates a 'barber shop' in her kitchen and sells cigarettes and candy to her customers. She does not have a local business licence and this money supplements her social assistance payments as a single mother. These business activities are an outgrowth of settlement life. What is important to recognize is that Native peoples, as they adapt to the ways of settlement life, are finding new ways to meet their needs. Whether or not these 'informal' economic activities are the forerunners of more viable and legal businesses, they do represent another facet of the ever-changing Native economy (Vignette 9.1).

The Land-Based Economy

The land-based economy consists of hunting, fishing, and gathering. There are nearly 100,000 trappers in the Canadian North and about half of these are Aboriginal peoples (Schellenberger and MacDougal 1986: 15). In the Arctic and Subarctic, many Native people continue to live in small, remote communities where alternative economic opportunities are limited, and where trapping remains an important source of income. While they welcome the technological advances which make their work easier, those born and raised in the bush are unlikely to abandon trapping for other occupations. One reason is that they enjoy the life, and another is that most lack the education and skills necessary for wage employment. Charlie Thomas, a trapper from the Old Crow Indian band illustrates these two points when he said (Schellenberger and MacDougal 1986: 16): 'I have been a trapper all my life. I had a family, but I lost my wife. I am just by myself now. I am 69 years old now, I draw the old age pension and I am still trapping. I have never been to school. I just live in the bush. All I know is how to sign my name on a cheque. I cannot read one word, but I make a living anyway.'

Trapping is a hard life with few monetary returns. Many young Natives have lost interest in a bush life, and Northwest Territories' fur data for a fifteen-year period show a downward trend in the number of pelts. Much of this decline is due to the collapse of the market for seal pelts (Table 9.7). The demise of the seal skin market was due to a European boycott that reflected pressure from animal rights groups. While they focused primarily on seals in the 1980s, animal rights groups have now widened the scope of their attack to include wild animal trapping. If the animal rights movement is successful in destroying markets for furs, then the northern trapping industry will, like the sealing industry, collapse. The low number of pelts sold in 1988/89 in the Northwest Territories may well signal such an impact. In spite of these drawbacks, trapping continues to be an extremely important source of cash to northern Aboriginal Canadians. In 1988/89, fur sales in the Northwest Territories amounted to $4.4 million.

When Native people lived on the land, trapping and hunting took place together. However, since Native people moved into settlements, hunting and fishing often take place separately. This change is but one more adaption to a new way of life. The new approach to obtaining country food has been successful. Country food is not only important for economic and nutritional reasons but it is also a key factor in preserving Native culture. Harvesting activities support the continuity of a land-based way of life, including Native cultural integrity and value system. The 'share

Table 9.7
Fur Production and Value, Northwest Territories, 1974/75 to 1988/89

Year	Total Pelts (No.)	($000)	Hair Seal Skins (No.)	($)
1974–75	230,629	2,082	34,565	596
1975–76	264,258	2,742	34,270	810
1976–77	416,759	4,317	48,407	822
1977–78	285,995	3,844	26,726	317
1978–79	214,070	5,738	29,352	416
1979–80	231,743	5,336	30,860	588
1980–81	272,915	5,026	42,120	890
1981–82	233,395	3,737	24,556	477
1982–83	190,293	2,794	14,837	221
1983–84	207,988	2,664	7,689	77
1984–85	163,176	3,241	5,419	54
1985–86	137,008	3,239	3,602	29
1986–87	178,071	5,623	2,563	33
1987–88	150,982	6,118	976	18
1988–89	69,788	4,405	1,696	33

SOURCE: GNWT Bureau of Statistics, Statistics Quarterly (1990), 12: 113(3).

ethic', for example, is a distinctive trait of Native culture and it provides a social mechanism for the distribution of wild game, birds, and fish among its members. The actual harvesting, preparing, and eating of game, fish, and other wild products from the land remains an intrinsic part of Native life. For instance, several Dene bands now undertake annual caribou hunts. The meat from these hunts is stored in community freezers and distributed to its members throughout the year. This adaptation of caribou hunting and sharing to modern circumstances bonds together band members and serves as a link between the past cultural practices and current ones. Country food is also very nutritious and raises the real cash income of Native people by reducing their purchases of store foods.

Economically, harvesting food from the land is important. During the mid-1980s, Usher and Wenzel (1989: 29) estimated that country food contributed around $55 million per year to the Native economy in the Northwest Territories. A Government of Northwest Territories report (1986: 10) estimated that the value of domestic consumption of game, fish, marine mammals, and other wildlife was just under $40 million in 1984/85. Usher and Wenzel suggested that, on average, each hunter is producing between $10,000 to $15,000 worth of food annually. Given that the 1985 per capita income of Aboriginal peoples age 15 years and over was $6,600 in Yukon and $5,660 in the Northwest Territories, the economic importance of country food is substantial. The estimated value of country food is a 'gross' figure and would be substantially lower if the

Table 9.8
Percentage of Persons Who Engaged in Hunting, Trapping, and Fishing, 1984

Region	On Land		Not On Land	
	Native	Non-Native	Native	Non-Native
Baffin	61	11	38	93
Kitikmeot	60	13	39	84
Keewatin	47	11	51	81
Inuvik	47	4	49	84
Fort Smith	28	2	70	96
Northwest Territories	46	3	52	93

SOURCE: Usher and Wenzel 1989: 55.

'costs' of securing this food were taken into account. In fact, an argument can be made that hunting and fishing are not economically sound activities, i.e., the cost of securing the game or fish exceeds the substitution value of store food. On the other hand, if Native people chose not to hunt and fish and if employment opportunities in Native settlements remained unchanged, then social welfare payments would rise sharply. Of course, there are powerful cultural reasons why some Native people continue to spend some time on the land and practice land-based activities. For these Native Canadians, the cultural and personal benefit of being on the land outweighs those benefits of remaining in the settlement year round.

Cultural Adjustment to New Institutions

Jobs, however, are but one issue facing Natives. While Native people have always shown an enormous capacity to adapt to new circumstances and to employ new technology to meet their needs, full-time employment in non-Native firms and agencies requires cultural adjustment. Examples of adaptation abound, ranging from the snowmobile to producing Inuit television programs. The land holds a special place in the hearts and minds of Native people. While they are not opposed to industrial development, many wish to participate in this process through their own economic institutions. In this way, Native people will have more control over the future of their people, institutions, and culture. Some Natives are already employed by private companies and public agencies, while others are employed by Native organizations. Given the number of Natives expected to seek work in the coming decade, many will have to work for non-Native firms.

A major barrier facing Native job seekers is limited education. Until this gap is closed, the Native labour force will not be able to compete with the non-Native one. Job training programs to prepare Native people for employment within businesses and government agencies have helped but until more teenage Native students graduate from high school most jobs will be closed to Native applicants. Training programs vary but each agency or company is attempting to provide basic skills to its potential employees. For example, the Department of Indian Affairs and Northern Development has been active in Native job training programs since 1973 when it established its On-The-Job Training Program. The Government of the Northwest Territories is attempting to increase the number of native employees in its civil service to approximate the proportion of Natives in the territorial population.

Federal Role in Native Economy

The dual provincial/federal political structure affects the delivery of northern programs. With a variety of delivery agencies, each responsible to a different government, sharp differences have arisen across the North. For example, the federal government has been the lead agent in the two territories while provincial governments have been the dominant force south of the 60th Parallel, except on Indian Reserves where the federal government's Department of Indian Affairs and Northern Development plays the key role. Generally, the federal government spends more money on its programs and is more innovative than its provincial counterparts. Usually, this means that the territories receive new programs first and that these programs are better funded than those in the provincial Norths.

Public housing programs provide almost all of the shelter for Native northerners. Few have sufficient income to afford modern housing and, without publicly built houses, many Natives would be living in shacks and log cabins. Federal and provincial governments have provided a variety of housing programs. In northern Saskatchewan, for example, government housing for the Métis is privately owned, though the provincial government with help from Central Mortgage and Housing Corporation provided the capital for house construction. Saskatchewan Housing holds the mortgages on these units and is responsible for collecting the monthly payments. The Métis are responsible for maintaining their houses. In contrast, housing on neighbouring reserves in northern Saskatchewan is owned by Indian bands, though the capital for their construction and maintenance comes from the Department of Indian Affairs and Northern Development.

In the two territories, the lead agency of the federal government is the Department of Indian Affairs and Northern Development (DIAND). This department is charged with the difficult task of co-ordinating northern development. The same department is also responsible for the well-being of native peoples. The dual mandate of this department leads to policy conflicts because the interests of developers do not necessarily coincide with those of Native people. For example, a basic assumption of DIAND is that resource development can benefit Native people through job creation and business opportunities. Low education levels for Aboriginal northerners make it difficult for them to take advantage of such developments.

Unsettled Business

In the early years of the northern resource boom, governments and developers ignored the interests of Native people. The Canol Project, the Alaska Highway, the Schefferville iron ore and railway project, Eldorado uranium mine, and other early examples of resource development projects eroded the Native land-base and, in doing so, threatened the very basis of Native culture. At that time, development took place as if local Natives did not exist. The James Bay hydro project dealt the Cree and Inuit of northern Quebec into the development equation and in this sense set a precedent for other Natives who now face the question of resource development on Indian or Inuit lands. Because the North still has large areas of Crown land, the geographic space necessary for a large land-base is possible in the two territories and, to a lesser degree, in the seven provinces. In southern Canada, the Crown no longer has large blocks of land and the cost of acquiring such land from private owners would be very expensive. For the Dene and Inuit, the prospective of establishing Native self-government within such land-bases seem favourable, though the powers of Native regional governments would likely fall far short of those associated with a province. Nevertheless, it could provide these Natives with a yet-to-be defined regional autonomy which, in conjunction with favourable cash settlements, may alter their place in Canadian society by allowing them to have more control over 'development'.

Vignette 9.2   Early Development and Natives

The story of resource development prior to the James Bay Project saw little benefit accrue to Natives who felt that outsiders were taking their land and resources and returning little to them. J.R. Miller, in a recent book, *Skyscrapers Hide the Heavens*, wrote that: 'Most of Canada's native peoples who were in the path of development found that they were ignored in the process of going after the resources or left out of the division of the proceeds of their sale. Aside from some relatively low-paid jobs in unskilled categories, the Cree (Chipewyans) of north-central Saskatchewan benefited little from Eldorado's Beaver Lodge mine. Still, they were luckier than the Lubicon band of central Alberta, who were not only ignored but seriously threatened by energy developments on lands that traditionally had been theirs. Though the Lubicon had much earlier been promised a reserve, the commitment was not honoured. And when an oil company began to disrupt their territories in the 1980s, their protests were ignored by a provincial government obsessed with economic development.'

Adapted from Miller 1989.

Dene, Métis, and Inuit living in the Canadian North have access to vast Crown lands which they perceive as 'their' land. As lands become attractive to southern-based companies, their status may change from Crown lands to leased or privately owned lands, where Natives are much less free to hunt and trap. In the early 1980s Usher (1982:187) stated:

> The traditional economy had weathered the changes of the post-war years to survive surprisingly intact and persisted in an uncertain accommodation with the growth of wage employment and government transfer payments. Thus, the traditional mode of production, relying on an enormous land base that could not possibly be reduced to small reserves, remained central to native life in the North.

As the federal government and Natives enter into the last land settlements, the hope is that these agreements will provide Native peoples with the land and cash base for participating in northern development on their terms. Not surprisingly, current land claims by northern Natives involve a large land base. It is this land base that Natives view as critical to their survival.

Aboriginal rights of Natives have been recognized in treaties. These treaties have established a 'social contract' between the members of the treaty and the federal government. The geographic expression of treaties is shown in Figure 9.1. Indians and Inuit without a treaty are in the process of securing such an agreement. Métis claim to a land base is still unclear, though the Alberta Government has awarded its Métis a land settlement.

Negotiations have been underway for some time and their settlement is not a simple matter. The federal government sees the agreement as falling in line with previous ones, while the Native leaders wish to link a form of self-government with the land settlement. On the one hand, there is an urgency to settle this unfinished business which would then allow Native peoples to focus their energies on 'development'. On the other hand, it is not certain that land claims settlements will improve their low economic standing in Canadian society.

Native Land Claims

Native land claims are a key element in Native-white relations. The legal basis for Native land claims in Canada is the concept of aboriginal rights: those rights stemming from the original use and occupancy of lands by Aboriginal peoples; they are also bound up with Native people's struggle for self-identity and self-determination. Section 35 of the Canadian Constitution (1982) affirmed the existence of aboriginal rights of Indian, Inuit, and Métis peoples of Canada but did not define these rights in precise legal terms. Until legal and/or political processes can determine their nature, aboriginal rights remain uncertain.

Vignette 9.3　The Basis of Treaties

At the time of Confederation, the British North America Act assigned the federal government responsibility for Indians and lands assigned for them. Later, when Canada turned its attention to the North, Inuit were to fall under federal jurisdiction. In 1982, the Canadian Constitution recognized Métis as one of the three Aboriginal peoples of Canada.

Figure 9.1
Treaty and Comprehensive Land Agreements

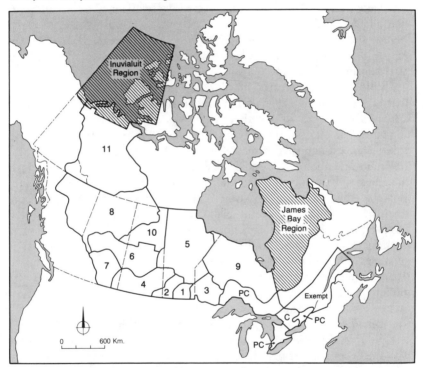

Indian Treaties are agreements between the Canadian government and Indian groups in which native people exchange their interest in specific areas of their ancestral lands for various payments and promises by the government. These agreements fall into three categories: pre-Confederation Treaties, Numbered Treaties, and Modern Treaties. By 1991, two modern (often called comprehensive) land claims had been settled: the James Bay Agreement (James Bay Cree in 1975, the Inuit of northern Quebec in 1975, and the Naskapis of northern Quebec in 1978) and the Inuvialuit Agreement in 1984. By the late 1980s, the claims of other northern Native groups, including the Yukon First Nations, the Tungavik Federation of Nunavut, and the Dene and Métis in the Northwest Territories, were well advanced. In 1991 the Gwich'in tribe located around the Mackenzie Delta broke away from the Dene / Métis claim and negotiated its own claim with the federal government. The Gwich'in Tribal Council, unlike others forming the Dene / Métis group, were prepared to surrender their aboriginal rights for the benefits of a comprehensive claim agreement. SOURCES: *The Canadian Encyclopedia* 1988: 1056–9 and McMillan 1988: 294.

The settlement and development of land in Canada was facilitated by treaties, agreements between the Crown and Indian people. The treaty involves an exchange between Native people and the British and later Canadian governments. Indians surrendered their rights to large land areas in return for various kinds of payments and promises from Crown officials. One of these promises was the right to hunt, fish, and trap on unoccupied Crown lands. Over time, the opportunity to hunt, fish, and trap has been reduced in much of the North as Crown land has been leased by logging companies or purchased by mining firms. Indian reserves were established for the use and benefit of bands which had taken treaty. (In Quebec and British Columbia, reserves were assigned to Indian tribes which had not taken treaty.) There are over 2200 reserves in Canada. Title to these lands is held by the Crown. Under the terms of the Indian Act, a band cannot sell or otherwise dispose of reserve lands without surrendering them to the Crown.

The first treaty occurred in 1763 at the close of the Seven Years' War. Treaty-making evolved over time, through three main phases: pre-Confederation Treaties, Numbered Treaties, and Modern Treaties. Modern treaty-making occurs in parts of Canada where Native rights have never been extinguished (except for the Indians living in the Mackenzie River Basin). Such treaty-making is referred to as comprehensive land claims as against specific land claims. Under the Indian Act, if a band feels that the terms of an existing treaty have not been fully discharged, then they can make a specific claim. Modern treaty-making is termed comprehensive because such negotiations deal with a wide range of issues: financial compensation, land ownership and use, hunting rights, extinguishment of aboriginal rights, and political rights. By 1991, three modern treaties had been signed. The 1975 James Bay Agreement was the first modern treaty. It involved Cree and Inuit peoples living in northern Quebec. In 1978, a similar treaty, the Northeastern Quebec Agreement, was signed with the Naskapi Indians living near Schefferville. The third one occurred in 1984 when the Inuvialuit people reached an agreement with the federal government. All agreements involved a cash settlement, land ownership, and a form of regional self-government. The James Bay Agreement may have set the terms for future modern treaties. The James Bay Agreement called for $225 million to be paid to the Cree and Inuit over a ten-year period; outright ownership of 13,300 km^2, and exclusive hunting rights over 155,000 km^2. In the 1984 Western Arctic Claim, the Inuvialuit will receive $45 million between 1984 and 1997, title to 11,000 km^2, exclusive hunting rights over 78,000 km^2, and rights to sand and gravel.

The land claim process must resolve the matter of land selection. Selection involves geographic concerns such as location and quality of land. Duerden (1990: 35–36) outlines four spatial components of the land claims process. These are:

(1) Mapping areas of use and occupancy in order to develop a case for legitimacy of a claim;

(2) Preparation of maps of land-use potential in order to identify 'optimum' areas to be retained as an outcome of negotiations;

(3) Evaluation of land selection positions as negotiation proceeds in order to establish the extent to which they satisfy the goals of the process;

(4) Evaluation of non-ownership agreements regarding land in order to ascertain the effectiveness of the control they offer and their long-term impact on land-use patterns.

The James Bay and Inuvialuit comprehensive land claims agreements have created a new set of Native corporations. In the case of the James Bay settlement, the Quebec Government put in place a guaranteed income system, designed to replace social assistance payments and to support the Native land-based economy.

Native Corporations

The most exciting prospect for northern development is Native investment of expected cash settlements associated with land claims. The experience of Alaskan Natives provides one model for investment strategies (Vignette 9.4). Coupled with the prospect of a substantial cash and land settlement, Native peoples are now more accustomed to settlement life and the market economy. Many young Native adults are better prepared for wage employment than ever before and there are indications that Native leaders have softened their attitude toward large-scale industrial projects. In a survey conducted in 1982, this change was strongly evident in the level of Dene support for the Norman Wells Oil Field Expansion and Pipeline Project. This grassroots support was not so much for the project, nor against the principle of 'no claims settlement, no development' espoused by the Dene Nation, but for the jobs and business opportunities that the project promised. Five years earlier, at the hearings of the Berger Inquiry, Native leaders and most Dene had spoken out strongly against the Mackenzie Valley Pipeline Project. What caused such a change over this short period of time? One factor may have been that high rates of unemployment and an inability to generate enough cash income from trapping led the average Dene to the conclusion that the overriding issue was 'jobs now'. Another factor may be related to the demographic change in the Dene population. From 1976 to 1982, the percentage of the Native adult population born and raised in settlements rose sharply, as did those with schooling. For these reasons young Dene are more at ease with and better able to enter the wage economy and participate in settlement life than are older Dene, whose formative years were spent hunting and trapping.

Vignette 9.4 Inuvialuit Development Corporation

Goals and Objectives:

The Inuvialuit Development Corporation is the business vehicle which will enable the Inuvialuit people to achieve a greater measure of self-sufficiency through involvement in sound economic and business ventures.

The overall goal of the Inuvialuit Development Corporation is to establish a stable, long-term economic base which will allow the Inuvialuit to contribute to and benefit from the regional and national economy. Some of the benefits to be gained include:

- economic growth through investment and profit
- Improvement of regional services
- greater long-term opportunities for employment and training locally and nationally
- a more stable economy through recycling of profits and wages in this region.

Strategy:

The Inuvialuit Development Corporation will achieve these objectives through the efficient deployment of its assets of more than $50 million. A portion of these assets are deployed in a combination of wholly owned and partly owned active businesses such as:

- Operation of Inuvialuit Corporate Centre in Inuvik and other commercial properties in Tuktoyaktuk
- Northern Transportation Company Limited (Marine Transportation in the North)
- Aklak Air Ltd. (Air Transportation in the Western Arctic)
- Grimshaw Trucking and Distributing Ltd. (Ground Transportation in the North)
- Stanton Distributing Co. Ltd. (Grocery Distribution and Catering Services in the Western Arctic)
- ATCO/Equtak Drilling Ltd. (Oilfield Drilling Contractor)
- Inuvialuit Sporting Goods Ltd. (Sporting Goods and Outdoor Supply Retailer in the Mackenzie Delta)
- Kerkoff Development Corporation (Real Estate Developer in British Columbia)
- Koblunaq Construction Ltd. (Real Estate Developer in Canada's North)
- Nutaaq Surveys Ltd. (Mapping & Survey Contractor in the North)
- Valgro Ltd. (Industrial Valve manufacturer for Western Canada)
- Arvik Environmental Services (Oil Spill Prevention in the Beaufort)
- Riverside Park Developments Inc. (Real Estate Developer on Vancouver Island)

The Inuvialuit Development Corporation is also actively pursuing other profitable ventures across Canada as a significant portion of our capital remains uncommitted.

SOURCE: *Arctic Circle* 1(1) (1990): 1.

In the post-settlement period, Native leaders too see an opportunity for a Native-controlled economy. Rather than participating as individuals in business, Native leaders see Native organizations operating their business ventures. This approach already exists at the band level and in joint ventures, but only in northern Quebec and in the western Arctic have land claims been settled and Native organizations obtained the land base and financial power to create a Native-controlled economy. This approach does not guarantee success, however. The troubled Alaskan Native corporations attest to business risks and the marginal nature of many Native enterprises (Robinson, Pretes, and Wuttunee 1989).

The total cash settlement of the Yukon, Dene, and Inuit claims could exceed $1 billion. With such capital resources, a different path to development could occur, in which Native corporations would provide the business structure for Native peoples, making investments on their behalf and employing Native workers. Some indication of the effect of a settlement on Native employment is found in the 1989 NWT Labor Force Survey. In this report, the Inuvialuit employment growth from 1984 to 1989 was the highest of the Native groups. At 46%, it was much higher than the Métis/Dene (31%) and the Inuit (17%). This suggests that the investments made by the Inuvialuit land-claims settlement have had a positive impact on local Native employment, although some of their investments are not in the Western Arctic (Vignette 9.4).

Given a limited physical base and market in the North, the key to economic advancement of Native people depends on participating in resource development. The success of such participation may well depend on the role of Native corporations in the 1990s and beyond. The unresolved question is whether Native corporations, with a wider scope of concerns than just a profit motivation, can survive in competition with private and public firms.

Staying on the Land: The Cree Income Security Program

The importance of traditional activities in the late 1980s is revealed by the fact that in 1988 43% of Native adults hunted and fished (Government of the Northwest Territories 1990 3:49). One of the most promising support programs for the native land-based economy is the Income Security Program for Cree Hunters and Trappers, arising out of the James Bay Agreement. It guarantees a minimum income for full-time Cree hunters and trappers. This program cost the Quebec Government about $6 million in 1981 (Salisbury 1986: 94) and nearly $12 million in 1985 (Ames et al. 1989: 275). Its rapidly increasing cost may be a critical reason why the federal government has not developed similar programs in the two territories. Other native organizations, such as the Dene Nation, have attempted to include this type of program in their land negotiations, so far without success.

Under the James Bay and Northern Quebec Agreement, the Quebec Government agreed to fund two programs designed to support traditional land-based Cree and Inuit activities. These programs are The Income Security Program for Cree Hunters and Trappers and the Northern Quebec Inuit Hunter Support Program. The rationale behind both programs was that public assistance to low-income groups is more effective in the form of a guaranteed income than as social assistance payments. For the Cree and the Inuit, these programs offered a means of supporting a traditional life-style by providing much needed cash income. The programs were designed to meet the needs of two distinct cultural groups, and the following discussion deals only with the Income Security Program for Cree Hunters and Trappers. The Cree program guarantees full-time hunters a minimum cash income each year plus an allowance for each day spent on the land. It ensures that hunting, fishing, and trapping remain a viable way of life for the Cree people, and that individual Cree who elect to pursue such a way of life are guaranteed a measure of economic security. The program encourages hunters and their families to stay on the land. In 1976–77, the first full year of this program, 1,012 families or single individuals registered, thereby indicating their intention to hunt all winter and, in return, receive a 'grub-stake' payment before departing for their hunting or trapping areas (Salisbury 1986:76).

Since then, a family or individual wishing to register with the program must in the previous year have spend more days in harvesting than in wage employment, and each person must have spent at least 120 person-days engaged in hunting, trapping, fishing, or gathering activities, 30 person-days of which can be spent in a settlement. Time spent purchasing supplies in the community or selling furs to the local fur buyer are considered as legitimate person-days. If a person does not spend 120 days on the land, then that individual is removed from the registry. Also, the basic payment is reduced as other sources of income, such as fur sales and wage earnings, rise. Fur sales below $575 per adult are not subject to this 40% reduction (see Table 9.9). In this way, trapping and casual employment are encouraged but the recipient only receives part of these earnings. Finally, people registered in the Income Security Program are not eligible for social assistance payments.

Cash payments to hunters are based on three factors–the size of the hunting party, their annual earnings, and the number of days spent on the land or engaged in related activities in a community. Each hunting party is assigned a basic cash payment and a per diem payment. In 1985, the basic payment was calculated as follows: for the hunter, $2298; for the spouse, $2298; family, $921; and each child, $921.

A single hunter would received $2298, while the basic payment for a family with two children would be $7,359. Each hunting party also receives a daily allowance for every person-day spent outside a permanent settlement while engaged in hunting, trapping, fishing or other country-food harvesting activities. The maximum number of days is set at 240; in 1985, the daily allowance was $27.15 for each adult. Both the basic payment

Table 9.9
Calculations of Income Security Payments in 1985

Per Diem Payment Hunter and Spouse (200 days × $27.15)			$10,860
Basic Payment			
Hunter	$2,298		
Spouse	2,298		
Family	921		
Child 1	921		
Child 2	921		
Total		$ 7,359	
Reduction of Basic Payment by Other Income			
Fur	1,100		
less $575/adult	−1,150		
	0		
Wages	4,500		
Per diem income	10,860		
Total Subject to 40% Reduction	15,360		
Reduction	−6,144	$ 1,215	
Total Benefits Paid			$12,075

SOURCE: Ames et al. 1989: 271.

and the daily allowance are indexed to the Quebec Pension Plan and they are adjusted upwards each year. The calculation of cash benefits for a family of four persons is shown in Table 9.9. In this example, a family of two adults and two dependent children received total cash benefits of $12,075 from the Income Security Plan, $4,500 from employment in the settlement, and $1,100 from fur sales. The family's total income was $17,675.

Summary

For centuries Native people have practised a distinct way of life in northern Canada. With the move to settlements, their land-based economy was no longer able to meet all their needs and wants. Native northerners now required a cash income to satisfy needs and wants common to all those living in an urban environment. The Native economy began to change, adapting to these new circumstances and opportunities. One such adaptation has been a much greater participation in the wage economy. Native employment in both the public and resource sectors is rising, although the average income of Native northerners remains far below that of non-Native northerners. Another adaptation to settlement life has resulted in changes in hunting and trapping. The traditional practice of the entire extended family staying on the land for most of the year has been replaced

---

Vignette 9.5   The Impact of the Income Security Program on Hunting by the Cree of Northern Quebec

Salisbury (1986: 76–84) describes changes in hunting among the Cree of northern Quebec. Comparing hunting in the pre-James Bay settlement time (1971) with a decade later, Salisbury stresses the increased availability of cash and therefore the ability to make more use of modern technology and transportation. In 1971, hunters took little equipment into the bush and often travelled only short distances. By 1981, the Cree hunters could make use of chartered aircraft to fly their families into remote bush camps which now were as comfortable as village housing. As well, hunting has become more mechanized and less physically exhausting because these Cree hunters can now afford to fly a snowmobile to their winter camps. The snowmobile is the work horse of the hunter. It allows a quick inspection of his trap line, and the easy hauling of big game and firewood from the bush to camp. Mechanization, made possible by the Income Security Program, has made hunting less strenuous and more productive. From 1974 to 1979, Salisbury reports that there was a 20% increase in the overall amount of game killed by these Cree hunters and their families (whose number increased by more than 50% over the same time period).

---

by short trips by male members of the group or family and by spring camping by the entire family at traditional hunting and fishing sites. An exception to this pattern occurs in northern Quebec where the Cree Income Security Program encourages Cree families to spend long periods of time on the land. Living in settlements has necessitated mechanized forms of transportation such as the snowmobile. Mechanization requires capital, forcing the Native trapper/hunter to seek government assistance and seasonal wage employment. Other adaptations have resulted in new institutional structures. One such structure is Native development corporations while another is air-commuting to Native communities provided by resource companies.

However, the most important changes for Native people are associated with comprehensive land agreements, available to those Native Canadians who have not yet entered into claims agreement with the Crown. Comprehensive agreements provide Native people with much more land, capital, and administrative control over their way of life than earlier treaties. The size of the cash and land settlement plus the level of political control allows for a measure of Native self-government within a region and, with that authority, the responsibility to spend/invest the cash settlement, to determine land-use policy, and to administer local services. Comprehensive land agreements provide a form of 'bottom-up' economic decision-making as well as the means for the Native economy to grow into a new, more viable form. Four immediate challenges face these new 'governments': (1) ensuring that there are sufficient jobs for their rapidly growing labour force, (2) resolving investment strategy trade-offs between secure 'external' investments or risky internal ones which would have the side-benefit of creating local jobs, (3) dealing with large-scale resource proposals such as the Beaufort oil and gas developments and the expansion of the James Bay Project, and (4) ensuring a healthy land-based economy where country-food harvesting is a sustaining one. In northern Quebec, income security programs provide a basic level of financial support living on the land but, as the Cree and Inuit populations continue to increase at very high rates, pressure on the wildlife resource may exceed its capacity to sustain itself. Added to this challenge is the expected loss of large hunting/fishing areas if the construction of the Great Whale River hydro-electric project proceeds.

In spite of these and other formidable economic challenges, regional Native governments must seek commercial solutions with private and public resource corporations and through this process exert control over the course of events affecting Native Canadians. This approach has the attraction of reducing dependency and rewarding regional initiatives, but it also carries with it the possibility of failure.

Suggested Readings

Bone, Robert M. and Milford B. Green, 1984. 'The Northern Native Labor Force: A Disadvantaged Work Force', *The Operational Geographer* 3: 12–14.

Duerden, Frank, 1990. 'The Geographer and Land Claims: A Critical Appraisal', *The Operational Geographer* 8(2): 35–7.

Peters, Evelyn J., 1989. 'Federal and Provincial Responsibilities for the Cree, Naskapi and Inuit Under the James Bay and Northern Quebec, and Northeastern Quebec Agreements', in *Aboriginal Peoples and Government Responsibility: Exploring Federal and Provincial Roles*, ed. by D.C. Hawkes. Chapter 6. Ottawa: Carleton University Press.

Robinson, Michael, Michael Pretes and Wanada Wuttunee, 1989. 'Investment Strategies for Northern Cash Windfalls: Learning from the Alaskan Experience', *Arctic* 42(3): 265–76.

Salisbury, Richard F., 1986. *A Homeland for the Cree: Regional Development in James Bay, 1971–1981*. Montreal: McGill-Queen's University Press.

Schellenberger, Stan and John A. MacDougall, 1986. *The Fur Issue: Cultural Continuity: Economic Opportunity*. Report of the House of Commons Standing Committee on Aboriginal Affairs and Northern Development. Ottawa: Queen's Printer.

Stabler, Jack C., 1989a. 'Dualism and Development in the Northwest Territories', *Economic Development and Cultural Change* 37(4): 805–40.

——, 1989b. 'Jobs, Leisure and Traditional Pursuits: Activities of Native Males in the Northwest Territories', *Polar Record* 25(155): 295–302.

Usher, Peter J., 1982. 'Unfinished Business on the Frontier', *The Canadian Geographer* 26(3): 187–90.

———— and George Wenzel, 1987. 'Native Harvest Surveys and Statistics: A Critique of Their Construction and Use', *Arctic* 40(2): 145–60.

———— and George Wenzel, 1989. 'Socio-Economic Aspects of Harvesting', Chapter 1 in *Keeping On the Land: A Study of the Feasibility of a Comprehensive Wildlife Harvest Support Program in the Northwest Territories* by Randy Ames, Don Axford, Peter Usher, Ed Weick, and George Wenzel. Ottawa: Canadian Arctic Resources Committee.

Wenzel, G.W., 1986. 'Canadian Inuit in a Mixed Economy: Thoughts on Seals, Snowmobiles, and Animal Rights', *Native Studies Review* 2(1): 69–82.

IV
SUMMING UP

Chapter 10

NORTHERN REALITIES

The Canadian North remains 'a region of conflicting goals, preferences and aspiration', reflecting its dominant but opposing images–a resource frontier and a Native homeland (Berger 1977: vii). While the images remain constant, since the Mackenzie Valley Pipeline Inquiry fundamental changes have created a new northern reality. Among these changes are (1) a series of mega-projects, built in six of the seven provincial 'norths' and the Northwest Territories, that have more fully integrated the North into the world economy, and (2) a shift in the balance of power between the major players (resource companies, governments, and Native organizations) largely because the Canadian public has become aware of the hidden cost of resource development and its adverse impact on the environment and on Native people. Multinational companies and Crown corporations that remain committed to 'profitable' resource development must now pay heed to environmental issues and to increasing Native political power. Nevertheless, the resource economy is the driving force behind northern development.

The North: A Modern Hinterland

Since the Second World War, the geography of the North has been the story of resource development. Over this time, the northern economy has shifted from a fur-oriented one to a resource hinterland economy; Native people have moved from the land to settlements; the environment has changed from a wilderness to one subject to resource development and pollution; and governments have extended southern institutions and programs into the North. During that fifty-year period, the resource economy has gone through two stages and is just entering a third. At first, resource development paid little heed to either the rights of Native people or the environment. Industrialists had a free hand to exploit natural resources and were often encouraged by various levels of government. By the late 1960s, a number of industry-caused environmental disasters surfaced and the public began to realize that there were hidden costs to resource development. One example was the discharge of mercury-charged effluent by a pulp plant into a river system which also served as the traditional fishing grounds of the Ojibway Indians. The damage done to the health of these Indians by this industrial pollution received world-wide attention. This tragic event demonstrated the ultimate consequences of toxic waste entering the food chain–the poisoning of humans.

By the 1970s, resource development had entered a second phase characterized by large-scale energy, mineral, and hydroelectric projects, by the establishment of environmental assessment agencies, and by a shift in public attitude–from approval of resource development to growing opposition. This opposition, while still a minority of Canadians in the late 1970s, gained substantial strength during the 1980s as more and more examples of environmental damage (such as the devastating impact of river diversions and vast reservoirs on the landscape), environmental pollution (such as the overflowing of a holding pond for uranium wastes), and 'unsound' resource practices (such as clear-cutting of timber over large areas) came to the public's attention. The resource industry had made itself the 'bad' guy in the development scenario. The Canadian public also began to connect major environmental impacts with adverse socio-economic impacts on Native people.

The second stage was characterized by more stringent environmental procedures and regulations. The third phase sees a growing interest in Canadian resources by those groups reportedly representing global environmental concerns. Led by the European Community, the seal industry has already suffered from a European boycott of its product and the fur industry is threatened by a similar type of boycott. This group has now put the Canadian forest industry on the defensive. They argue that Canadians are blessed with one of the few remaining 'great' boreal forests and must treat it as a trust for all peoples. To Europeans, this means a halt to Canadian clear-cut logging practices or a boycott of Canadian forest products. Canadian forest industry spokespersons argue that clear-cutting is a more efficient means of logging, thereby keeping costs down and exports up. Yet selective cutting, while more expensive, does employ more loggers and might encourage the Canadian forest industry to process more wood products (again resulting in more employment). Another sign of the globalization of environmental issues is the June 1991 Arctic Environmental Project Strategy declaration, in which Canada and seven other arctic nations acknowledge the international nature of arctic pollution and the need for an integrated approach to arctic environmental issues. Cleaning up the Canadian Arctic is an expensive proposition, and the continuing deposit of toxic particles on the land and in the seas complicates efforts to eliminate these contaminants from the arctic food chain. The federal government recently announced it was committing $100 million to an arctic clean-up and studies to identify how toxic chemicals from distant factories and mines enter the arctic food chain (Platiel 1991: A7). All of this means environmental clean-up and research jobs, many of which may well involve northern residents.

Choices and Trade-offs

In the last decade of the twentieth century, Canadians are faced with critical choices and trade-offs. For example, should more resource projects be encouraged, or should such lands be set aside for wilderness parks and for Native land claims agreements? While Canadians wish to protect the environment and make generous land claims settlements, more jobs are required to meet the needs of the growing labour force (particularly in Native settlements) and more tax revenue is needed to pay for environment agencies, to provide subsidies to resource companies and transfer payments to the territories, and to honour land claims settlements.

Canadians face hard choices involving trade-offs between the environment and resource development, and Native employment or social welfare. What are the prospects for job creation in the North? The North's future may well depend on three economic choices available to all peripheral economies (Britton 1988: 156–7). All choices assume that 'development' is the desired economic state and that it is market-driven. They also imply that a 'development' ladder exists; i.e., there is the opportunity to progress from a resource economy to a diversified one. Such assumptions are more likely to come to fruition in the Subarctic than in the Arctic because its resource base is more varied and it has better access to markets. Other factors differentiating the two regions are the non-renewable nature of the Arctic's resource base and its small Inuit labour force which has little experience with a market economy.

The continued exploitation of the region's resources and their export to world markets remains the first choice. This approach calls for a progression over time from primary production to a more diversified economy where resource processing plays a prominent role. Such diversification reduces the problem of export vulnerability; that is, a downturn in the demand for export products could send the regional economy into a tailspin–a version of the 'boom-bust cycle' associated with primary economies. This approach dominates Canadian political thought and underlies the basic northern strategy of provinces and territories. Alberta's pulp mills and Quebec's hydroelectric projects are two examples. The Northern Accord between the federal government and the territories may be seen as another instrument for the same strategy. This agreement proposes to the territories exclusive control over on-shore resources and the tax revenues generated by such resource development, and a share of off-shore resources. As the Premier of the Northwest Territories commented, 'The Northern Accord is the big one upon which all our political hopes ride' (Fisher 1988: A3). In effect, revenues from the Beaufort oil and gas could make the Northwest Territories a 'have' territory and allow jobs to be created in the renewable resource sector. Beaufort oil and gas development could trigger an economic boom in the Northwest Territories and generate resource royalties, but economic diversification through business spin-offs from the oil boom is unlikely because small northern firms cannot compete with large companies based in southern Canada. The experience of the Norman Wells Project demonstrated both a high rate of economic leakage during construction and few business opportunities for northern firms in the operational phase of the project. Since there is no reason to believe future mega-projects will function differently, they offer insufficient economic stimulation to diversify the northern economy. Oil royalties from off-shore energy development are not a sure thing–at least not in the foreseeable future. While additional energy revenues will no doubt flow to the government of the Northwest Territories once a northern energy accord is signed, the share of off-shore oil revenue going to the territorial government is problematic. The federal government may reduce transfer payments proportionally to increases in oil royalties, leaving the territorial government no better off financially than before.

The second choice is to encourage tourist industries, trading on the North's spectacular scenery and its history. This economic strategy will broaden the hinterland's economy and diversify employment opportunities, particularly in remote communities. One weakness is the short northern summer. The seasonal peak in tourism creates two problems for the North. First of all, employment is seasonal, and secondly the costs of such industries must be met by summer revenues, making the price to the consumer very high and less competitive with alternative recreational experiences in southern Canada. A second problem is caused by tourists who engage in fishing and hunting. These tourists, by reducing the stock of fish and game, may be adversely affecting the subsistence activities of Native people.

The third choice is for hinterland economies to develop new export products based on human capital, especially among the Native population. Eskimo carving business represents a past success and more innovations along these lines should be attempted, but such cottage industries often fail because of the difficulty of creating a commercial market and ensuring a regular supply. Other avenues to explore deal with the environment. One need is to create a workforce to clean up existing pollution; another is to study the causes of environmental pollution. The challenge of building pipelines in the Arctic where continuous permafrost exists and where pipelines must cross large stretches of the Arctic Ocean, offers exciting research opportunities which could be jointly funded by oil and gas companies and interested governments. A scientific establishment at an arctic centre like Inuvik (perhaps expanding the existing federal Scientific Resource Centre) would provide ready access to suitable continuous permafrost environment near the Mackenzie Delta and Beaufort oil and gas deposits. Such a scientific centre might also receive both manpower and financial support from the Inuvialuit Development Corporation.

While these development strategies offer some prospect for job creation and economic growth, a large segment of the northern population–those Native adults living in small, remote communities–will be unaffected. For this reason, the Native economy deserves another look. Some fifteen years ago, Berger (1977[I]: xxvi) stated:

> I am convinced that non-renewable resources need not necessarily be the sole basis of the northern economy in the future. We should not place absolute faith in any model of development requiring large-scale technology. The development of the whole renewable resource sector–including the strengthening of the native economy– would enable native people to enter the industrial system without becoming completely dependent on it.
>
> An economy based on modernization of hunting, fishing and trapping, on efficient game and fisheries management, on small-scale enterprise, and on the orderly development of gas and oil resources over a period of years–this is not retreat into the past; rather, it is a rational program for northern development based on the ideals and aspirations of northern native peoples.
>
> To develop a diversified economy will take time. It will be tedious, not glamorous, work. No quick and easy fortunes will be made. There will be failures. The economy will not necessarily attract the interest of the multinational corporations. It will be regarded by many as a step backward. But the evidence I have heard has led me to the conclusion that such a program is the only one that makes sense.

Native Economy

For Native northerners, the global market economy has not yet fulfilled its promise of jobs and a better life. Given the location of most Native people in remote areas, job opportunities may be expected to continue to elude Native job-seekers. With the continuing decline of the trapping industry, the alternative Native economy offers few commercial opportunities. While the younger, more educated Native northerners are likely to find employment in Native organizations, the resource industry, or government agencies, others, particularly those with little schooling, may wish to stay on the land but need cash support. The Income Security Programs for the Cree and Inuit of northern Quebec offer two models of subsidized hunting. This approach is based on the rationale that it is better to be productively engaged in land-based activities than drawing social assistance. The program guarantees a minimum cash income for full-time hunters and provides an economic base for those wishing to continue to pursue a hunting/trapping life style. At the same time, it reduces the competition for jobs in the settlements. Its only drawback is the possibility that children of hunting families may not opt for an education fitting them for wage employment.

The New Map of the North

Comprehensive land agreements and the proposed territorial division will change the political face of the North, transferring power to Native people through their regional governments and offering them new opportunities. Already, the James Bay and Inuvialuit agreements have changed the political map of the North. The James Bay and Inuvialuit agreements have resulted in Native regional government, Native development corporations, and a new sense of direction and purpose. The James Bay and Northern Quebec Agreement in 1975 was Canada's first major modern land-claim settlement. The second settlement was reached between the federal government and the Committee for Original People's Entitlement in 1984. These two land-claims settlements resulted in Native title to vast areas of northern Quebec and the western Arctic. More recently, the Yukon, Dene / Métis, and Inuit land claims have been agreed to in principle. These settlements alter the political landscape of northern Canada and offer the promise of Native self-government. As well, each of these agreements involves large cash settlements, providing capital for investments by the various Native corporations representing each of the Aboriginal peoples.

The proposed division of the Northwest Territories by the Dene and Inuit represents another major change of the map of Canada's North. Originally proposed by the Inuit Tapirisat of Canada in 1976, this proposal is close to approval. The main stumbling block is the failure of the two Native organizations to reach an agreement on the territorial boundary. Once in place, the Inuit would be better able to control their political destiny, though they would still require large transfer payments from Ottawa to operate Nunavut. The acceptance of such a division will lead to a form of 'home' rule for the Inuit. The situation is less clear for the Dene because these Native people would then form a minority in their

Table 10.1 Land Claim Settlements, to 1991

CLAIM	AGREEMENT TYPE AND DATE	NUMBER OF BENEFICIARIES AT SIGNING	$ COMPENSATION TOTAL / PER CAP.	OTHER PAYMENTS	ONGOING INCOME	LAND SURFACE IN TOTAL SQUARE MILES / PER CAP.	SURFACE WITH SUBSURFACE TOTAL / PER CAP.	EXCLUSIVE HUNTING, FISHING, AND TRAPPING LANDS / PER CAP.
JAMES BAY CREES	FINAL NOV. 11, 1975 LEGISLATED LATER	6,650	$135 MILLION / $29,300	$145 MILLION REMEDIAL. APPROX. $17.5 MILLION 'CATCH-UP' FUNDS FOR SERVICES	HUNTERS AND TRAPPERS FAMILY SUPPORT, FIGURES NOT AVAILABLE	2,140 sq. mi. / 0.32 sq. mi.	— / —	24,899 sq. mi. / 3.74 sq. mi.
INUIT OF NORTHERN QUEBEC	FINAL NOV. 11, 1975 LEGISLATED LATER	4,390	$90 MILLION / $20,501	$12.6 MILLION 'CATCH-UP' FUNDS FOR SERVICE	HUNTERS AND TRAPPERS COMMUNITY SUPPORT. FIGURES NOT AVAILABLE	3,147 sq. mi. / 0.72 sq. mi.	— / —	33,631 sq. mi. / 7.66 sq. mi.
NASKAPIS OF N.E. QUEBEC	FINAL AGREEMENT JAN. 12, 1978	390	$9 MILLION / $23,136	$1.26 MILLION 'CATCH-UP' FUNDS FOR SERVICES.	HUNTERS AND TRAPPERS FAMILY SUPPORT. FIGURES NOT AVAILABLE	270 sq. mi. / 0.69 sq. mi.	— / —	1,600 sq. mi. / 4.10 sq. mi.
INUVIALUIT	FINAL AGREEMENT JUNE 5, 1984	2,500	$55 MILLION / $22,000	$7.5 MILLION SOCIAL FUND. $16.8 MILLION IMPLEMENTATION.	POTENTIAL GRAVEL ROYALTY. POTENTIAL OIL AND GAS ROYALTY.	30,000 sq. mi. / 12.5 sq. mi.	5,000 / 1.56 sq. mi	6,000 sq. mi. YUKON NORTH SLOPE / 2.5 sq. mi.
INDIANS OF YUKON	UMBRELLA FINAL AGREEMENT MAR. 31, 1990	6,500	$242.7 MILLION / $37,340	$11 MILLION PREPARATION AND IMPLEMENTATION.	SHARE OF GOV'T ROYALTY. POTENTIAL ROYALTIES	6,000 sq. mi. / 0.92 sq. mi.	10,060 sq. mi. / 1.54 sq. mi.	IN SMALL PARKS, ETC.
DENE/MÉTIS OF N.W.T.	FINAL AGREEMENT APRIL 9, 1990	13,000	$500 MILLION / $38,461	$21 MILLION FOR NORMAN WELLS IMPACT, DISCUSSED PRIOR TO AGREEMENT.	SHARE OF GOV'T ROYALTY. POTENTIAL ROYALTIES.	66,100 sq. mi. / 5.08 sq. mi.	3,900 sq. mi. / 0.3 sq. mi.	7,500 sq. mi. WOOD BUFFALO PARK / 0.6 sq. mi.
INUIT OF N.W.T.	AGREEMENT IN PRINCIPLE APRIL 30, 1990	17,000	$580 MILLION / $34,118		SHARE OF GOV'T ROYALTY (HUNTER'S SUPPORT UNDER NEGOTIATION). POTENTIAL ROYALTIES.	121,360 sq. mi. / 7.14 sq. mi.	14,000 sq. mi. / 0.82 sq. mi.	—

SOURCE: K.J. Crowe, 'Claims on the Land', *Arctic Circle* 1 (3) (1990): 20–1.

Figure 10.1
Nunavut, and Its Changing Boundaries

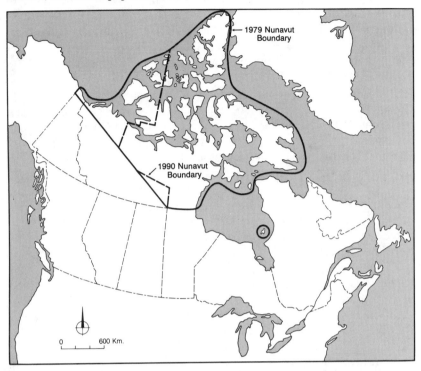

The original boundaries of Nunavut changed as the Inuvialuit claimed the Western Arctic and the Dene disputed the border along the Barrenlands. The Gwich'in Indians of the Mackenzie Delta reached a land-claim settlement in July 1991. The tribe will receive ownership of nearly 24,000 km² of land straddling the border with Yukon. This agreement-in-principle was ratified by the Gwich'in Tribal Council in September 1991. SOURCE: Wonders 1987: 669.

section of the Northwest Territories, thereby losing much of the political power now held by Natives in the legislature of the Northwest Territories (Dacks 1986).

Nunavut may well follow the development path of Greenland which has had home rule since 1979. Under home rule, Greenland remains part of Denmark but has legislative powers similar to a province, i.e., Denmark retains control of the constitution, citizenship, external affairs, defence, monetary policy, and the justice system. Most significantly, most of the Greenland national budget is derived not from local taxes but from funds supplied by Denmark. During the past decade, the Greenland economy has grown steadily, though negative growth did occur in 1981/82 (Lyck 1989: 345). Oddly enough, unemployment has risen rapidly and the explanation is that many Greenlanders have insufficient education and training to obtain employment. In a pattern similar to that of the Northwest

Territories, Danes continue to hold a disproportionate number of jobs, particularly the higher paying professional and management positions. In this sense, Greenland, like the Northwest Territories, is still dependent on the non-Native labour force for most of its trained workers. Political semi-independence, whether it is called Home Rule or Native self-government, does not alter the fundamental education and training shortcomings found in the Native labour force.

The Greenland experience is replicated in most of the Canadian North. Northern Quebec seems to be an exception, partly because regional government has created many job opportunities for Cree people while others can take advantage of the Hunters Support Program. Salisbury (1986: 131) described education as the key to the emergence of Cree regionalism. The signing of the James Bay agreement in 1975 created a need for educated Cree to staff the Cree Regional Authority, the Cree Regional School Board, and the large number of organizations formed under the agreement. It was not until the early 1980s that the number of Cree high school graduates began to match the vacant positions with the Cree bureaucracy (Salisbury 1986: 127). Regional Cree-controlled government and a more positive employment outlook dramatically changed students' attitude toward school–now it had meaning in their lives and the lives of other Cree.

In other parts of the Canadian North, the drop-out rates for Native students remains high. In the Northwest Territories, only 5% of Native students who start school graduate from high school (Government of the Northwest Territories 1989: 36). Since almost all permanent jobs in industry and government require *at least* a high school diploma, the vast majority of the Aboriginal workforce are not qualified for such positions. Waiving such educational requirements for Native job-seekers may not be in their best long-term interests; Native workers lacking basic skills may have trouble keeping a job or being promoted. Failure to resolve the school drop-out problem will insure continued dependency on government and inability to find employment. As the Scone Report (Government of the Northwest Territories 1989: 69) stresses: 'Our survival as peoples of the North–whether our ancestors lived here for centuries or whether we arrived in recent years–has always depended upon our ability to educate ourselves, organize ourselves, and live a life of personal and community discipline.'

To the Year 2000

With the liberalization of world trade, the role of the North in the global economy will remain a resource hinterland, supplying energy, minerals, and timber to world markets. Under the Free Trade Agreement with the United States, more resource development is likely to be tied to the American market. On the other hand, diversification programs funded by federal, provincial, and territorial governments will be more difficult because subsidies for industrial plants and preferential treatment to local firms (an import substitution policy) can no longer be used to stimulate regional economic growth. Efforts by provincial governments to reduce trade barriers within Canada will add to the liberalization of trade within Canada. Without trade barriers, regional economic development will be determined by each region's comparative advantage–which leaves the North as a primary producer.

Within the North, resource development will continue to take place primarily in the Subarctic, postponing exploitation of the vast arctic oil and gas deposits to the next century. Among the difficulties of exploiting arctic resources is cost, The Arctic is the highest cost area in Canada in which to undertake exploration, construction, and operation of mineral deposits and petroleum fields. Moreover, the Arctic contains a limited mix of resources and therefore its economic potential is restricted to a narrow resource base. Its major commercial resources are non-renewable; long-term economic stability based on these resources is therefore very doubtful. With the exception of Beaufort oil and gas, the Arctic's prospects for economic growth in the coming decade are much less promising than those for the Subarctic. In the case of its vast oil and gas fields, arctic resource development first requires pipeline research studies in a cold environment, and then vast sums to construct these pipelines. Such mega-projects are usually undertaken by multinational companies and Crown corporations. But until the world price for these products rises sharply, even these large companies will shy away from arctic oil and gas development.

Resource developments in the Subarctic may well accelerate in the next ten years. Most resource development is likely to take place in the provincial Subarctic and existing operations such as coal production in British Columbia, pulp manufacturing in Alberta, uranium mining in Saskatchewan, hydroelectric power exports in Manitoba and Quebec, iron ore output in Labrador, and a mix of forest and mining operations in Ontario could benefit from higher prices as the world economy emerges from its recession.

While there is no doubt that large-scale resource development increases the value of mineral and energy production in the North, it is also true that the number of permanent jobs generated is small. Since mineral and energy products are exported, labour-intensive jobs associated with processing these raw materials are located in other regions and countries, limiting the impact of mega-projects on the northern economy. The counter argument is that there is no alternative to mineral and energy resource development. For these reasons, territorial and provincial governments look to resource development as a means of stimulating economic growth and, by creating more jobs, as a partial solution for social and economic problems affecting their northern areas. In the case of most provinces, northern resource development has a very positive impact on the provincial economy. In the case of Quebec, the economic well-being of the province is affected by the James Bay and other massive hydroelectric projects. The provincial strategy of economic development is based heavily on hydroelectric power and the export of this power to southern markets, particularly those markets in New England. The construction of new dams and hydro facilities is expected to boost the Quebec economy and much of the increase in revenue will be passed on by Hydro Quebec to the provincial government. This strategy is based on two important assumptions: (1) that the New England demand for electrical power will provide a market for surplus energy, and (2) that as new units come on stream the hydroelectric system will generate substantially more power than that required by Quebec consumers.

While the Cree, Inuit, and Naskapi peoples of northern Quebec obtained substantial benefits from the James Bay Agreement, the cost of this agreement is seen in the flooding of hunting and trapping lands. The 10,000 or so northern Native Quebec residents received land and a cash settlement of $225 million. While there is some argument over whether or not this was a 'fair' settlement, it did give these Aboriginal peoples more control over their lives and capital to invest in their own firms. In 1990, for example, the Makivik Corporation, an Inuit business firm created after the James Bay Agreement, has had its offer to purchase controlling interest in First Air, the largest air carrier in the Northwest Territories, accepted. This scale of Native business purchases and operations could not have occurred prior to the James Bay settlement and, while there are no guarantees that all Native enterprises will succeed in the competitive market place, the rise of Native entrepreneurs within the framework of Native corporations augurs well for both their economic and cultural future. By using their cash settlement from land claims to invest in the North, a new era of entrepreneurship for Native development corporations has arisen, creating job opportunities for Native persons and generating much greater multiplier effects than similar investments by outside firms. Coupled with the prospect of more land-claim settlements by other Native groups, the northern economy is destroying the myth that Native people do not want to participate in the wage economy and operate businesses. As Native peoples become more accustomed to settlement life and the market economy, their involvement is likely to increase.

Such trade-offs are changing the human geography of the North. Non-Native people will encourage more careful treatment of the environment and increased employment of the Native workforce (Vignette 10.1). While the northern economy will remain a resource hinterland for the foreseeable future and its 'modernization' will continue, economic, political and economic power is shifting to some northern Natives through comprehensive land claims. What they do with that power over the next decade or so may well break the dependency ties with government, and help determine the shape and direction of Aboriginally-controlled development in the next century.

Vignette 10.1 Gold Mine To Give Preference To Indians

North of Pickle Lake in the remote Dona Lake area of Northwestern Ontario, a gold company (Placer Dome) and a union (United Steelworkers of America) are trying to agree on an employment-equity plan for native people that could have far-reaching implications. The union proposes that when the rights of native employees conflict with the rights of non-native employees, 'the rights of native employees . . . shall prevail.' In all cases of hiring, promotion, transfer, layoff and recall from layoff, native employees would be given preferential treatment if they have the ability and the physical fitness to perform the work. The Steelworkers' lawyer said that 'for far too long, natural resource extraction that has occurred near or on native reserves has failed to provide the natives with employment benefits related to that extraction. Historically, companies mining in native areas didn't employ any natives, and they liked it that way.' Placer Dome's Dona Lake mining operation stands as a notable exception, with 26 native employees in a work force of 126. In 1987, the company signed a native employment equity agreement with the Osnaburgh Band, the Windigo Tribal Council, and the provincial and federal governments. In exchange, the native groups dropped their request for an environmental assessment of the Dona Lake gold mine.

SOURCE: Galt 1991: A4.

Selected Readings

Berger, Thomas R., 1977. *Northern Frontier, Northern Homeland: The Report of the Mackenzie Valley Pipeline Inquiry*, vols. 1 and 2. Ottawa: Department of Supply and Services.

Dacks, Gurston, 1986. 'The Case Against Dividing the Northwest Territories', *Canadian Public Policy* 12 (1): 202–13.

Government of the Northwest Territories, 1989. *The Scone Report: Building Our Economic Future*. Yellowknife: Legislative Assembly of the Northwest Territories.

Salisbury, Richard E., 1986. *A Homeland for the Cree: Regional Development in James Bay, 1971–1981*. Montreal: McGill-Queen's University Press.

Appendices

Appendix 1.1
The Method of Calculating Nordicity

Nordicity is a quantitative measure of ten variables found at a particular northern place or site. Each variable is subjectively assigned a value (called polar units by Hamelin). The sum of these ten variables for a place give a measure of its nordicity. For instance, the maximum number of polar units theoretically attainable is 1,000. For two variables (types of ice and population), there are several options and the selection of the 'best' option is determined by the particularly location of the site. The boundary between the North and the rest of Canada has a value of 200 polar units. A summary for the calculation of nordicity by each variable is shown below. For more information, see Hamelin (1979 and 1988) in the references for Chapter 1.

Variable		Polar Units
1. Latitude	90°	100
	80°	77
	50°	33
	45°	0
2. Summer Heat	0 days above 5.6°C	100
	60 days above 5.6°C	70
	100 days above 5.6°C	45
	>150 days above 5.6°C	0
3. Annual Cold	6650 degree days below 0°C	100
	4700 degree days below 0°C	75
	1950 degree days below 0°C	30
	550 degree days below 0°C	0
4. Types of Ice		
Frozen Ground		
Continuous permafrost 457 m thick		100
Continuous permafrost <475 m thick		80
Discontinuous permafrost		60
Ground frozen for less than 1 month		0
Floating Ice		
Permanent pack ice		100
Pack ice for 6 months		36
Pack ice <1 month		0

Glaciers
    Ice sheet >1523 m thick   100
    Ice cap 304 m thick   60
    Snow cover <2.5 cm   0

5. Annual Precipitation
    100 mm   100
    300 mm   60
    500 mm   0

6. Natural Vegetation cover
    Rocky desert   100
    50% tundra   90
    Open woodland   40
    Dense forest   0

7. Accessibility by land or sea
    No service   100
    For two months   60
    By two means   15
    > two means   0

8. Accessibility by Air
    Charter, 1600 km   100
    Regular service, weekly   25
    Regular service, daily   0

9. Population
  Settlement size
    None   100
    About 100   85
    About 1000   60
    >5000   0

  Population density
    Uninhabited   100
    1 person per km^2 50
    4 persons per km^2 0

10. Degree of economic activity
    No production   100
    Exploration   80
    20 hunters/trappers   75
    Interregional Centre   0

Appendix 2
Population Change by Census Divisions 1981 and 1986

	1981	1986	% Change
Yukon (01)	*23,153*	*23,504*	*1.5*
Northwest Territories:	*45,741*	*52,238*	*14.2*
Baffin Region (04)	8,300	9,975	20.2
Fort Smith Region (06)	22,344	25,116	12.4
Inuvik Region (07)	7,485	8,411	12.4
Keewatin Region (05)	4,327	4,986	15.2
Kitikmeot Region (o8)	3,285	3,750	14.2
British Columbia:	*227,556*	*225,590*	*−0.1*
Bulkley-Nechako District (51)	38,309	37,470	−2.2
Fraser-Fort George District (53)	89,431	89,337	−0.1
Kitimat-Stikine District (49)	42,400	39,483	−6.9
Peace River-Liard District (55)	55,463	57,278	3.3
Stikine Region (57)	1,953	2,022	3.5
Alberta:	*87,636*	*97,631*	*11.4*
Division 16	43,573	48,779	11.9
Division 17	44,063	48,852	10.9
Saskatchewan:	*25,304*	*25,340*	*0.1*
Division 18	25,304	25,340	0.1
Manitoba:	*73,668*	*73,996*	*0.1*
Division 19	12,277	9,125	−25.7
Division 21	24,714	24,068	−2.6
Division 22	26,673	30,544	14.5
Division 23	10,004	10,259	2.5
Ontario:	*443,826*	*434,060*	*−2.2*
Algoma District (57)	133,533	131,841	−1.3
Cochrane District (56)	96,875	93,712	−3.3
Kenora District (60)	59,421	52,834	−11.1
Thunder Bay District (58)	153,997	155,673	1.1
Quebec:	*487,943*	*473,326*	*−3.0*
Abitibi (84)	93,529	94,410	−3.0
Chicoutimi (94)	174,441	174,625	0.1
Lac-Saint-Jean-Ouest (90)	62,952	62,977	0.0
Saguenay (97)	115,881	104,131	−10.1
Territoire-du-Nouveau-Québec (98)	41,140	37,183	−9.6
Newfoundland:	*57,056*	*54,695*	*−4.1*
Division 9	25,738	25,954	−4.1
Division 10	31,318	28,741	−8.2
Northern Census Division Total	1,471,883	1,460,380	−0.1

SOURCE: Statistics Canada. 1987. Population: Census Divisions and Subdivisions. Catalogue 92-101. Ottawa: Minister of Supply and Services.

Appendix 3
Major Urban Centres by Northern Census Divisions, 1986

Northern Census Divisions	Major Urban Centre	1986 Pop.	Percentage Change
Yukon	Whitehorse	15,199	2.6
Northwest Territories:			
Baffin Region	Frobisher Bay	2,947	26.3
Fort Smith Region	Yellowknife	11,753	23.9
Inuvik Region	Inuvik	3,389	7.7
Keewatin Region	Rankin Inlet	1,374	23.9
Kitikmeot Region	Cambridge Bay	1,002	22.9
British Columbia:			
Kitimat-Stikine Regional District (49)	Kitimat	11,196	−12.6
Bulkley-Nechako Regional District (51)	Smithers	4,713	3.0
Fraser-Fort George Regional District (53)	Prince George	67,621	0.1
Peace River-Liard Regional District (55)	Fort St John	13,355	−3.9
Stikine Region (57)	Stikine SRD	1,750	1.6
Alberta:			
16	Fort McMurray	34,949	12.7
17	Slave Lake	5,429	20.5
Saskatchewan:			
18	La Ronge	2,696	4.5
Manitoba:			
19	Berens River Reserve	803	17.9
21	The Pas	6,283	−1.7
22	Thompson	14,701	2.9
23	Leaf Rapids	1,956	−17.2
Ontario:			
Cochrane (56)	Timmins	46,657	1.2
Algoma (57)	Sault Ste Marie	80,905	−22.0
Thunder Bay (58)	Thunder Bay	112,272	−0.2
Kenora (60)	Kenora	9,621	−2.0
Quebec:			
Abitibi (84)	Val-d'Or	22,252	3.8
Lac-Saint-Jean-Ouest (90)	Roberval	11,448	0.5
Chicoutimi (94)	Chicoutimi	61,083	1.5
Saguenay (97)	Baie-Comeau	26,244	−2.3
Territoire-du-Nouveau-Québec (98)	Chibougamau	9,922	−7.5
Newfoundland:			
9	St Anthony	3,182	2.4
10	Labrador City	8,664	−24.9

SOURCE: Statistics Canada, 1987/88. *Profiles: Census Divisions and Subdivisions.* Catalogue 94-101 to 124. Ottawa: Minister of Supply and Services.

Glossary

ABORIGINAL PEOPLES: Those Canadians of Indian, Inuit, and Métis ancestry.

ACCESSIBILITY: The ease with which interchange can occur between two or more places. One measure of accessibility is the cost of travel or communications between places.

ASSIMILATION: The process of integrating one culture into another so that the former losses its distinctiveness.

BIRTH RATE: The annual number of births per 1000 individuals within a given population.

CIRCUMPOLAR WORLD: Those seven countries in the northern hemisphere that contain either Arctic or Subarctic environments: Canada, Finland, Iceland, Norway, Sweden, Union of Soviet Socialist Republics, and United States of America.

CORE/PERIPHERY MODEL (Heartland/Hinterland model): A general theory of polarized economic growth which results in an industrial region and a resource-oriented one, and in which decisions and demand in the industrial region shape the economic direction of the surrounding hinterland. This concept has both a Marxist and a conventional economic interpretation. It is applicable at all geographic scales: internationally, nationally, or regionally. In this book, the model is applied to the North (a resource periphery) and the industrial world (core).

COUNTRY FOOD: Food obtained by hunting, fishing, or gathering. While country food is no longer the chief source of food for many Native northerners, it remains an important and popular food for Native families, particularly those in more remote communities. Most studies have concluded that country food is more nutritious than store food and that it is important culturally to Native people.

COUREURS DE BOIS: Unlicensed fur traders of New France who played an important role in the European exploration of Canada and in establishing trading contacts with the Indians.

CRYOSOLS: Thin soils formed in the continuous permafrost zone. These soils have active or thawed layers less than one metre thick.

DEATH RATE (Also called crude death rate): The annual number of deaths per 1000 population in a given population.

DENE: Indian people of Canada's western Subarctic. Traditionally, the Dene lived by hunting, fishing, snaring wild animals, fish and birds, and gathering berries.

DENENDEH: Native self-determination in the Northwest Territories called for an autonomous state within Canada: Denendeh. The Dene Nation attempted to tie this political concept to its land claims settlement but the federal government, while later accepting the idea of dividing the Northwest Territories roughly along the treeline, would not allow this linkage.

DEPENDENCY: A corollary of dominance; a situation where a region or people must rely on other regions for their economic well-being. Dependency can also mean that external capital and technology play a paramount role in the regional economy and that local politicians have little power in the provincial and federal governments.

DIFFUSION: The spread or movement of a concept, practice, article, or population from a point of origin to other areas.

DRUMLIN: Streamlined elongated hill composed of glacial deposits; its long axis parallels the direction of glacier flow.

DUAL ECONOMY: Two types of social and economic systems existing simultaneously within the same territory; in the case of the Canadian North, the resource economy and the native economy.

ECONOMIES OF SCALE: The cost advantage gained by large-scale productions; these arise as the average cost of production falls within increasing output.

ECUMENE: Permanently inhabited areas of the earth.

ESKER: Ridge of sand and gravel, sometimes many kilometres long, deposited by streams beneath or within a glacier.

ETHNOCENTRISM: The perception that one's own ethnic group is superior to other ethnic groups.

EXTENDED FAMILY: Social unit comprising parents and children and other more distant relatives.

FERTILITY RATE: The number of live births in a given year per 1000 women aged 15-44.

FJORD: Steep-walled glacial trough that has been invaded by the sea, producing a deep inlet; found along mountainous coasts in high latitudes.

FOURTH WORLD: Encapsulated or enclaved Native societies in economically advanced countries. In this book, the term refers to the Canadian North as a place where the original inhabitants have become minorities — and generally disadvantaged minorities — within their 'own' lands. What defines Fourth World peoples, but especially northern Native people, is their utilization of land as a source of food, spiritual strength, and a common resource base.

FURAN: Colourless, volatile, liquid hydrocarbon (C_4H_4O) used in the synthesis of organic compounds.

GELIFLUCTION (Solifluction): The movement of thawed, wet soil downslope in a series of distinct lobes; occurs in tundra regions.

GLACIAL TILL: Unsorted and unstratified material carried and laid down by glacial ice.

GREENHOUSE EFFECT: Atmospheric warming by transmission of incoming short-wave solar radiation and trapping of outgoing longwave terrestrial radiation.

GROSS NATIONAL PRODUCT: The total value of all goods and services produced by a country per year.

GROWTH RATE: The rate at which a population is increasing (or decreasing) in a given year due to natural increases and net migration; expressed as a percentage of the base population.

HINTERLAND: The market area or region served by an urban centre; see core/periphery model.

HOMEGUARD: Indian employed at a trading post.

HOMELAND: Local natural features, cultural traits and economic issues, the basis of commonality that provides a distinctive regional personality. The image held by native northerners of the North.

HUMAN ECOLOGY: The application of ecological concepts to the study of the relations between people and their physical and social environment.

ICE FOG: Smog-like feature found in larger urban centres such as Whitehorse and Yellowknife. Ice fog occurs during prolonged cold spells and inversions.

INDIAN: The term Indian is a legal one in Canada. A person whose name is on the band list of any Indian community in Canada, or on the central registry list in Ottawa, is an Indian. The main Indian linguistic groups in the Canadian North are Algonkian and Athapaskan.

INFANT MORTALITY RATE: The number of deaths of infants under one year of age in a given year per 1000 live births in that year.

INFORMAL ECONOMY: That part of an economy performing productive, useful, and necessary labour without a formal system of control and remuneration and operating beyond official recognition.

INVERSION: An atmospheric condition in which cold air underlies warm air; inversions are highly stable conditions and thus not conducive to atmospheric mixing.

ISOLINE: A map line connecting points of constant value, such as pressure (isobar) or temperature (isotherm).

ISOSTATIC UPLIFT: Isostacy is a state of balance maintained in the crust of the earth. Disturbance of the balance causes isostatic movements to act to restore the balance. Movements may involve uplift of the land to compensate for erosion, or depression of the land to accommodate the weight of accumulated ice or sediment.

LABOUR FORCE/WORKFORCE: The economically active population, consisting of productive employed and temporarily unemployed people.

LIFE EXPECTANCY: The average number of additional years a person would live if current mortality trends were to continue. Most commonly cited as life expectancy at birth.

MEGA-PROJECTS: Large-scale industrial undertakings, which, because of their enormous size, dominate the local and regional economy during the construction phase. Construction costs usually exceed $1 billion.

MÉTIS: During the early fur trade era many fur traders took Indian wives. Their offspring, the Métis, developed a separate culture and history. In 1982, the Métis gained official recognition as one of the three Aboriginal peoples of Canada.

METROPOLIS: A chief city of a country or region; the capital. Commonly has a population exceeding $1/2$ million.

MORAINE: Any of several types of landforms composed of debris transported and deposited by a glacier.

MORBIDITY: The frequency of disease and illness in a population.

MORTALITY RATE: Refers to deaths as a component of population change. The rate depends on age, sex, race, occupation, and social class. Its incidence can reveal much about a population's standard of living and health care.

MULTINATIONAL (company or corporation): A business organization usually head-quartered in one parent country but with established operations in several countries.

MUSKEG: An Indian term for a sphagnum-moss-covered bog in the subarctic. It is a region of marshy depressions, stagnant, mosquito-infested pools and slow meandering rivers.

NATIVE PEOPLE (Aboriginal Peoples): Those Canadians of Indian, Inuit, Métis, and Non-Status Indian ancestry.

NON-STATUS: Those Canadian Indians who by birth, marriage, or choice have no legal status, under the Indian Act, to benefit from reserve lands and special federal programs.

NORDICITY: A measure of the degree of 'northernness' of a place. Concept created by Hamelin. Nordicity provides a quantitative definition of the southern boundary of the North. It also allows a composite measure of northernness for any place. Nordicity is based on ten selected variables which are supposed to represent all facets of the North.

NUCLEAR FAMILY: Social unit comprising a man and a woman living together with their children.

NUNAVUT: 'Our land' in Inuit dialects found in the eastern Northwest Territories. A proposal for a new territory and government to be carved from the present Northwest Territories. Despite government support, Nunavut has been delayed by a series of boundary disputes with the Dene/Métis. The disputes are caused by the overlap of traditional hunting land-use by northern Indians, particularly the Chipewyan who as hunters of caribou seasonally occupied much of the forest-tundra transition zone. This dispute is complicated by the fact that several Chipewyan bands are now settled in northern Saskatchewan and Manitoba and have no direct voice in the deliberations.

OPEC (Organization of Petroleum Exporting Companies): The 13 major oil-exporting countries of the Third World acting as a cartel or oligopoly to promote their joint national interests. Members include Saudi Arabia, Nigeria, Algeria, Venezuela, Libya, Kuwait, United Arab Emirates, Iran, Iraq, Ecuador, Qatar, Gabon, and Indonesia.

OZONE DEPLETION: Ozone is the main atmospheric gas that absorbs biologically damaging ultraviolet radiation, and there are concerns that chlorofluorocarbons (CFCs) are depleting the ozone layer. An Antarctic ozone hole was discovered and since 1979 has become increasingly larger. In the Arctic, a craterlike hole was discovered in 1986 but does not appear to be increasing in size.

PCBs (polychlorinated biphenyls): Industrial insulating material.

PATTERNED GROUND: Ground in which vegetation, water features, or stones are arranged in a geometric pattern, e.g., circles or polygons. It is widespread in cold environments.

PEMMICAN: A food made by plains Indians; dry meat (usually buffalo) pounded into a coarse powder and then mixed with melted fat and possibly dried Saskatoon berries. It was light, simple, and did not spoil.

PERMAFROST: A ground condition in which the soil or subsoil is permanently frozen; long-term frozen ground in periglacial environments; ground remaining at or below the freezing point for at least two years.

PINGO: Domed hill formed when soil cover is pushed up by a lens-shaped mass of ice; occurs in tundra regions.

PODSOL: One of the soil groups under the traditional USDA Soil Classification schemes; a leached soil containing oxides of iron, formed mainly in cool, humid climates.

POLAR WORLD: Those lands with Arctic and Subarctic natural environments.

POLYNYA ('north water'): Large area of open water surrounded by sea ice; typical of the northern part of Baffin Bay.

POPULATION PYRAMID: A diagrammatic representation of the age and sex structure of a population.

POPULATION TRAP (Malthus): Belief that populations, if unchecked, tend to increase at a geometric rate while subsistence increases at an arithmetric rate. The trap occurs when a country or region has a higher rate of population increase than its rate of economic growth; eventually people are forced to live at a subsistence level.

PRIMARY ACTIVITIES: Economic pursuits involving production of natural or culturally improved resources, such as agriculture, livestock raising, forestry, fishing, and mining.

PSYCHIC INCOME: Nonmonetary rewards — such as pleasure or satisfaction — gained from choosing a lower-paying economic option.

PUSH-PULL MIGRATION: A theory used to explain the movement of people from rural to urban centres. The migrants are forced out of one area by limited opportunity and attracted to cities by perceived advantages.

QUALITY-OF-LIFE INDEX: The psychological, individual aspects of social well-being.

REGION: A portion of the earth that has some internal feature of cohesion or uniformity; for example, the Arctic.

RESERVE: Lands granted by the Federal government when treaties were made with the Natives. The purpose was to protect the Natives from settlers and prospectors and to establish a Native land base. Ownership of land granted to the tribe or band was a communal arrangement and land could not be sold. There are over 2200 reserves in Canada and title to the land is held by the Crown.

SCRIP: Scrip (a form of promissory note) was given to the Métis people and could be converted either into cash or land. For the most part scrips were sold, which resulted in the loss of a 'reserve' land base.

SOLIFLUCTATION/SOLIFLUCTION (Gelifluction): The movement of thawed, wet soil downslope in a series of distinct lobes; occurs in tundra regions.

STATUS (Non-Status): Canadian Indians who have 'status' fall under the Indian Act. They have rights to use reserve lands held by their band, and access to federal funding for programs such as housing and education. Those status Indians whose ancestors signed a treaty also have treaty rights.

SUBSISTENCE SYSTEMS: A system of relatively simple technology in which people produce most or all of the goods to satisfy their own and their family's needs; little or no exchange occurs outside of the immediate or extended family.

SUSTAINABLE DEVELOPMENT: Development that meets the needs of the present without compromising the ability of future generations to meet their own needs, and that advocates movement towards local control of production for local needs.

TERTIARY ACTIVITIES: The services and commerce portion of an economy. Examples include repair and maintenance of capital goods, personal services, public administration, medical care, transport and communication, etc.

THERMOKARST: Ground-surface depression created by the thawing of ground ice in the periglacial zone.

THRESHOLD: The minimum market needed for a business to survive or to justify the establishment of a public agency in a community.

TRIBAL GROUPS: Groups of people united by language and customs and belonging to the same band.

TUNDRA: A vegetational zone of lichen, mosses, sedges, and dwarf trees in high latitudes.

URBAN PRIMACY: Concentration of population and services in a region's largest city; cities of intermediate size are rare, and there are many very small settlements.

URBANIZATION: The process of becoming urban. Urbanization is associated with the concentration of population into towns and cities; city formation and expansion.

References

Chapter 1

Alonso, William, 1980. 'Five Bell Shaped Curves', *Papers of the Regional Science Association* 45: 5-16.

Asch, Michael, 1984. *Home and Native Land: Aboriginal Rights and the Canadian Constitution*. Toronto: Methuen.

Berger, Thomas R., 1977. *Northern Frontier, Northern Homeland: The Report of the Mackenzie Valley Pipeline Inquiry*. 2 vols. Ottawa: Minister of Supply and Services.

Bone, Robert M. and Milford Green, 1987. 'Frontier Dualism in Northern Saskatchewan', *The Operational Geographer* 12: 21-4.

Bourne, L.S. and J.W. Simmons (eds), 1978. *Systems of Cities: Readings on Structure, Growth and Policy*. Toronto: Oxford University Press.

Bradbury, John H., 1979. 'Towards an Alternative Theory of Resource-based Town Development in Canada', *Economic Geography* 55(2): 147-66.

Burns, B.M., F.A. Richardson, and C.N.H. Hall, 1975. 'A Nordicity Index', *The Musk-Ox* 17: 41-4.

Cruikshank, J., 1977. 'Myths and futures in the Yukon Territory: The Inquiry as a Social Dragnet', *Canadian Issues* 2 (2): 53-69.

Dacks, Gurston, 1981. *A Choice of Futures: Politics in the Canadian North*. Toronto: Methuen.

de Souza, Anthony R., 1986. 'To Have and To Have Not: Colonialism and Core-Periphery Relations', *Focus* 36 (3): 14-19.

Diubaldo, Richard J., 1984. 'The North in Canadian History: An Outline', *The Journal of Polar Studies*, 1: 187-96.

Dunbar, Moira, 1966. 'The Arctic Setting', in *The Arctic Frontier*, edited by R.St J. Macdonald. Toronto: University of Toronto Press.

Dyck, N., 1985. 'Aboriginal Peoples and Nation-States: An Introduction to the Analytical Issues', in *Indigenous Peoples and Nation-States: 'Fourth-World' Politics in Canada, Australia, and Norway*. St John's: Institute of Social and Economic Research, Memorial University.

Fellmann, Jerome, Arthur Getis, and Judith Getis, 1990. *Human Geography: Landscapes of Human Activities*. Dubuque: Brown.

Frank, A.G., 1969. *Capitalism and Underdevelopment in Latin America: Historical Studies of Chile and Brazil*. New York: Monthly Review Press.

Friedmann, John, 1972. 'A Generalized Theory of Polarized Development', in *Growth Centres in Economic Development*, edited by N. Hansen. New York: The Free Press.

Gamble, D.J., 1986. 'Crushing of Cultures: Western Applied Science in Northern Societies', *Arctic*, 39(1): 20-3.

Government of Canada, Department of Indian and Northern Affairs, 1989. *Transition* 2(8): 4.

Government of the Northwest Territories, Bureau of Statistics, 1989. *1989 NWT Labor Force Survey, Winter 1989: Overall Results and Community Detail*. Yellowknife: Department of Culture and Communications.

Grant, Shelagh D., 1988. *Sovereignty or Security: Government Policy in the Canadian North 1936-1950*. Vancouver: University of British Columbia Press.

————, 1989. 'Myths of the North in the Canadian Ethos', *The Northern Review* 3/4: 15-41.

Hamelin, Louis-Edmond, 1972a. 'L'ecoumene du Nord Canadien', in *The North: Studies in Canadian Geography*, edited by William C. Wonders. Toronto: University of Toronto Press: 25-40.

————, 1972b. 'A Zonal System of Allowances for Northern Workers: An Example of Applied Geography', *The Musk-Ox* 10: 5-20.

————, 1979. *Canadian Nordicity: It's Your North Too*. Translated by William Barr. Montreal: Harvest House Ltd.

————, 1988a. 'Nordicity', in *The Canadian Encyclopedia*, edited by James H. Marsh. Edmonton: Hurtig Publishers, 3: 1504-06.

————, 1988b. *About Canada: The Canadian North and Its Conceptual Referents*. Canadian Studies Directorate. Ottawa: Minister of Supply and Services.

Hearne, Samuel, 1958. *A Journey from Prince of Wales's Fort in Hudson's Bay to the Northern Ocean*, edited by Richard Glover. Toronto: Macmillan.

Ironside, R. Geoffrey, 1990. 'Regional Development Aid in the Peripheral Region of Northern Alberta', in P.J. Smith and E.L. Jackson (eds) *A World of Real Places*. Edmonton: Department of Geography.

Lane, Theodore, 1986. 'An Essay on the Meaning and Consequences of Economic Development.' Paper presented at the Western Regional Science Association Meetings at Laguna Beach, California in February 1986.

Lotz, Jim, 1970. *Northern Realities: The Future of Northern Development in Canada*. Toronto: New Press.

McCann, L.D., ed., 1987. *A Geography of Canada: Heartland and Hinterland*, 2nd ed. (Scarborough: Prentice-Hall)

Markusen, Ann R., 1987. *Profit Cycles, Oligopoly, and Regional Development.* Cambridge: The MIT Press.

Michalak, W.A. and K.J. Fairbairn 1988. 'Producer Services in a Peripheral Economy', *Canadian Journal of Regional Science* 11 (3): 353-72.

Morton, W.L., 1972. *The Canadian Identity*, 2nd ed. Toronto: University of Toronto.

Muller-Wille, L. and P. Pelto, 1979. 'Political Expressions in The Northern Fourth World: Inuit, Cree, Sami', *Etude Inuit Studies* 3(2): 5-17.

Myrdal, Gunnar, 1957. *Economic Theory and Underdeveloped Regions*. London: Methuen.

Oppong, J.R. and R.G. Ironside, 1987. 'Growth Centre Policy and the Quality of Life', *The Canadian Geographer* 10(3): 281-300.

Page, Robert, 1986. *Northern Development: The Canadian Dilemma*. Toronto: McClelland and Stewart Ltd.

Pretes, Michael, 1987. 'Underdevelopment in Two Norths: The Brazilian Amazon and the Canadian Arctic', *Arctic* 41(2): 109-16.

Rostow, W.W., 1961. *The Stages of Economic Growth: A Non-Communist Manifesto*. Cambridge: Cambridge University Press.

Rowley, Graham, 1978. 'Canada: the Slow Retreat of "The North"', in Armstrong, Terence, George Rodgers and Graham Rowley, *The Circumpolar North: A Political and Economic Geography of the Arctic and Subarctic*. London: Methuen.

Salisbury, Richard F., 1986. *A Homeland for the Cree: Regional Development in James Bay, 1971-1981*. Montreal: McGill-Queen's University Press.

Santos, Milton, 1979. *The Shared Space*. London: Methuen.

Stabler, Jack C., 1989. 'Dualism and Development in the Northwest Territories', *Economic Development and Cultural Change* 37(4): 805-40.

Shkilnyk, Anastasia M., 1985. *A Poison Stronger than Love: The Destruction of an Ojibwa Community*. New Haven: Yale University Press.

Stohr, W.B., 1981. 'Development from Below: the Bottom-up and Periphery-inward Development Paradigm', in W.B. Stohr and D.R.F. Taylor (eds) *Development from Above or Below? The Dialectics of Regional Planning in Developing Countries*. Chichester: John Wiley.

Todaro, Michael P., 1989. *Economic Development in the Third World*. New York: Longman.

Usher, Peter J., 1982. 'The North: Metropolitan Frontier, Native Homeland?' in McCann, L.D., *Heartland and Hinterland: A Geography of Canada*, 2nd ed. Scarborough: Prentice-Hall.

————, 1987. 'The North: One Land, Two Ways of Life', in McCann, L.D., *Heartland and Hinterland: A Geography of Canada*, 2nd ed. Scarborough: Prentice Hall.

Weissling, Lee E., 1989. 'Arctic Canada and Zambia: A Comparison of Development Processes in the Fourth and Third Worlds', *Arctic* 42(3): 208-16.

Williams, Peter J., 1986. *Pipeline & Permafrost: Science in a Cold Climate*. Ottawa: Carleton University Press.

Zaslow, Morris, 1988. *The Northward Expansion of Canada, 1914-1967*. Toronto: McClelland and Stewart.

Chapter 2

Barr, William, 1972. 'Hudson Bay: The Shape of Things to Come', *The Musk-Ox* 11:64.

————, 1986. *Glacio-Isostatic Rebound and the Changing Face of the Northwest Passage*. Paper presented to the joint meeting of the Hakluyt Society and the Society for the History of Discoveries, Brown University, Providence, R.I. in October 1986.

Bird, J.B., 1972. 'The Physical Characteristics of Northern Canada', in *The North: Studies in Canadian Geography*, edited by William C. Wonders. Toronto: University of Toronto Press.

Bostock, H.S., 1964. 'Physiographic Subdivisions of Canada', in *Geology and Economic Minerals of Canada* edited by R.J.W. Douglas. Ottawa: Queen's Printer.

Burns, C.R. and A.G. Lewkowicz, 1990. 'Retrogressive Thaw Slumps', *The Canadian Geographer*, 34(3): 273-6.

Dyke, Arthur S. and Thomas F. Morris, 1988. 'Drumlin Fields, Dispersal Trains, and Ice Streams in Arctic Canada', *The Canadian Geographer* 32 (1): 86-90.

Etkin, Dave, 1989. 'Elevation as a Climate Control in Yukon', *Inversion: The Newsletter of Arctic Climate* 2: 12-4.

French, H.M., 1976. *The Periglacial Environment*. London: Longman.

Graf, William L., ed. 1987. *Geomorphic Systems of North America*. Centennial Special Volume 2. Boulder: The Geological Society of America Inc.

Haggett, Peter, 1983. *Geography: A Modern Synthesis*. London: Harper & Row.

Hare, F. Kenneth and Morley K. Thomas, 1979. *Climate Canada*, 2nd ed. Toronto: Wiley.

Heginbottom, J.A. (co-ordinator), 1989. 'A Survey of Geomorphic Processes in Canada', in *Quaternary Geology of Canada and Greenland*, ed. R.J. Fulton. Ottawa: Minister of Supply and Services.

Ives, Jack D. and Roger G. Barry, 1974. *Arctic and Alpine Environments*. London: Methuen.

Johnson, P.G., 1988. 'Rock glaciers, Southwest Yukon', *The Canadian Geographer* 32(3): 277-80.

Lawford, R.G., 1988. 'Climatic Variability and the Hydrological Cycle in the Canadian North: Knowns and Unknowns', *Proceedings of the Third Meeting on Northern Climate*. Canadian Climate Program. Ottawa: Minister of Supply and Services: 143-62.

Laycock, A.H., 1987. 'The Amount of Canadian Water and Its Distribution', in *Canadian Aquatic Resources*, edited by M.C. Healey and R.R. Wallace. Ottawa: Department of Fisheries and Oceans: 13-42.

Mackay, J. Ross, 1989. 'Ice-Wedge Cracks, Western Arctic Coast', *The Canadian Geographer* 33 (4): 365-8.

Sater, John E., 1969. *The Arctic Basin*. Washington: The Arctic Institute of North America.

Slaymaker, Olav, 1988. 'Physiographic Regions', in *The Canadian Encyclopedia* edited by James H. Marsh, 2nd ed. Edmonton: Hurtig. 3: 1671.

Sugden, David, 1982. *Arctic and Antarctic: A Modern Geographical Synthesis*. Oxford: Blackwell.

Trenhaile, Alan S., 1990. *The Geomorphology of Canada*. Toronto: Oxford University Press.

Washburn, A.L., 1979. *Geocryology: A Survey of Periglacial Processes and Environments*. Norwich: Arnold.

Williams, Peter J., 1979. *Pipelines and Permafrost: Physical Geography and Development in the Circumpolar North*. London: Longman.

———, 1986. *Pipelines & Permafrost: Science in a Cold Climate*. Ottawa: Carleton University Press.

Woo, Ming-Ko, 1985. 'Focus: Hydrology of Snow and Ice', *The Canadian Geographer* 29 (2): 173-83.

Young, Steven B., 1989. *To the Arctic: An Introduction to the Far Northern World*. New York: Wiley.

Chapter 3

Abele, Frances, 1989. *Gathering Strength*. Komatik Series, Number 1. Calgary: The Arctic Institute of North America.

Anderson, R.M., 1924. 'The Present Status and Future Prospects of the Larger Mammals of Canada', *Scottish Geographical Magazine* 40(6): 321-31.

Bone, Robert M., Earl Shannon, and Stuart Raby, 1973. *The Chipewyan of the Stony Rapids Region*. Mawdsley Memoir 1. Saskatoon: Institute for Northern Studies, University of Saskatchewan.

Bryan, Alan Lyle, 1986. 'The Prehistory of Canadian Indians', in *Native Peoples:*

The Canadian Experience, edited by R. Bruce Morrison and C. Roderick Wilson. Toronto: McClelland and Stewart, Chapter 3.

Clark, Donald W., 1983. 'Mackenzie — River to Nowhere', *The Musk-Ox* 33: 1-9.

Cooke, Alan and Clive Holland, 1978. *The Exploration of Northern Canada: 500 to 1920, A Chronology*. Toronto: The Arctic History Press.

Crowe, Keith J., 1974. *A History of Original Peoples of Northern Canada*. Montreal: Arctic Institute of North American and McGill-Queen's University Press.

Damas, David, 1968. 'The Eskimo' in *Science, History and Hudson Bay*, edited by C.S. Beals. Vol. 1. Ottawa: Queen's Printer.

Donaldson-Yarmey, 1989. 'Alberta's First Fort', *Western People*, 8 June: 10.

Duffy, R. Quinn, 1988. *The Road to Nunavut: The Progress of the Eastern Arctic Inuit since the Second World War*. Kingston: McGill-Queen's University Press.

Estabrook, Barry, 1982. 'Bone Age Man', *Equinox*, 1(2): 84-96.

Fagan, Brian M., 1987. *The Great Journey: The Peopling of Ancient America*. London: Thames and Hudson.

Fumoleau, Rene, 1976. *As Long As This Land Shall Last: A History of Treaty 8 and Treaty 11, 1870-1939*. Toronto: McClelland and Stewart.

Gajda, Roman, 1960. 'The Canadian Ecumene — Inhabited and Uninhabited Areas', *Geographic Bulletin*, 15: 5-18.

Gordon, Bryan C. 1981. 'Man-Environment Relationships in Barrenland Prehistory', *The Musk-Ox* 28: 1-19.

Harris, R. Cole, 1987. *Historical Atlas of Canada. Vol. I (From the Beginning to 1800)*. Toronto: University of Toronto Press.

Hearne, Samuel, 1958. *A Journey From Prince of Wales's Fort in Hudson's Bay to the Northern Ocean: 1769-1772*. Toronto: Macmillan.

Hickey, Clifford G., 1986. 'The Archaeology of Arctic Canada', in *Native Peoples: The Canadian Experience*, edited by R. Bruce Morrison and C. Roderick Wilson. Toronto: McClelland and Stewart, Chapter 4.

Hilliard, Sam B., 1987. 'A Robust New Nation, 1783-1820', in *North America: the Historical Geography of a Changing Continent*, edited by Robert D. Mitchell and Paul A. Groves. Totowa: Rowman & Littlefield: 149-71.

Irving, W.N., 1968. 'The Barren Grounds', in *Science, History and Hudson Bay*, edited by C.S. Beals. Volume 1. Ottawa: Queen's Printer: 26-54.

Krech, Shepard, III, 1978. 'On the Aboriginal Population of the Kutchin', *Arctic Anthropology*, 15 (1): 89-104.

MacDonald, Jack, 1990. 'Daishowa Firm Seeks Police Protection', *The Edmonton Journal*, 28 September: A7.

McGhee, Robert, 1975. 'The Peopling of Arctic North America', in *Arctic and Alpine Environments*, edited by Jack D. Ives and Roger G. Barry. London: Methuen. Chapter 15: 831-55.

————, 1989. *Ancient Canada*. Ottawa: Canadian Museum of Civilization.

McMillan, Alan D., 1988. *Native Peoples and Cultures of Canada*. Vancouver: Douglas & McIntyre.

Miller, J.R., 1989. *Skyscrapers Hide the Heavens: A History of Indian-White Relations in Canada*. Toronto: University of Toronto Press.

Morantz, Toby, 1984. 'Economic and Social Accommodations of the James Bay Inlanders to the Fur Trade', in *The Subarctic Fur Trade: Native Social and Economic Adaptations*, edited by Shepard Krech III. Vancouver: University of British Columbia Press: 55-79.

Morrison, William R., 1983. *A Survey of the History and Claims of the Native Peoples of Northern Canada*. Ottawa: Department of Indian Affairs and Northern Development.

Morton, Arthur S., 1973. *A History of the Canadian West to 1870-71*, 2nd ed. Toronto: University of Toronto Press.

Ray, Arthur J., 1974. *Indians in the Fur Trade: Their Role as Trappers, Hunters, and Middlemen in the Lands Southwest of Hudson Bay, 1660-1870*. Toronto: University of Toronto Press.

————, 1984. 'Periodic Shortages, Native Welfare, and the Hudson's Bay Company, 1670-1930', in *The Subarctic Fur Trade: Native Social and Economic Adaptations*, edited by Shepard Krech III. Vancouver: University of British Columbia Press: 1-20.

————, 1990. *The Canadian Fur Trade in the Industrial Age*. Toronto: University of Toronto Press.

Rich, E.E., 1967. *History of the Hudson's Bay Company, 1670-1870*. Vol. II. London: Hudson's Bay Record Society.

Rohmer, Richard, 1970. *The Green North*. Toronto: Maclean-Hunter.

Rowley, Graham, 1978. 'Canada: The Slow Retreat of "the North" '. in *The Circumpolar North* by Terence Armstrong, George Rogers, and Graham Rowley. London: Methuen.

Star-Phoenix, 1989. 'Lubicon Band Patrols Oilfields on Disputed Land', 2 December: D14.

Sugden, David, 1982. *Arctic and Antarctic: A Modern Geographical Synthesis*. Oxford: Basil Blackwell.

Tough, Frank, 1988. 'The Northern Fur Trade: A Review of Conceptual and Methodological Problems', *The Musk-Ox* 30: 66-79.

Trudel, Marcel, 1988. 'Jacques Cartier', in *The Canadian Encyclopedia*. Edmonton: Hurtig Publishers.

Vorsey, Louis de, 1987. 'The New Land: The Discovery and Exploration of Eastern North America', in *North America: The Historical Geography of a Changing Continent*, edited by Robert D. Mitchell and Paul A. Groves. Totowa: Rowman & Littlefield.

Watson, J. Wreford, 1964. *North America: Its Countries and Regions*. London: Longmans.

Wolfe, Eric R., 1982. *Europe and the People Without History*. Berkeley: University of California Press.

Zaslow, Morris, 1984. *The Northwest Territories, 1905-1980*. Canadian Historical Association Historical Booklet No. 38. Ottawa: Canadian Historical Society.

————, 1988. *The Northward Expansion of Canada, 1914-1967*. Toronto: McClelland and Stewart.

Chapter 4

Berry, B.J.L., 1961. 'City Size Distribution and Development', *Economic Development and Cultural Change* 9: 673-87.

Bone, Robert M., 1972. 'The Population of Northern Canada', in *The North: Studies in Canadian Geography*, edited by William C. Wonders. Toronto: University of Toronto Press.

Choinière, Robert and Norbert Robitaille, 1987. 'The Fertility of the Inuit of Northern Quebec: A Half-Century of Fluctuations', *Acta Borealia* 1/2: 53-64.

Crowe, Keith J., 1974. *A History of the Original Peoples of Northern Canada*. Montreal: McGill-Queen's University Press.

El-Shakhs, S., 1972. 'Development, Primacy and the System of Cities', *Journal of Developing Areas* 7: 11-36.

Gajda, R., 1960. 'The Canadian Ecumene Inhabited and Uninhabited Areas', *Geographic Bulletin* 15: 5-18.

Hagey, N. Janet, Gilles Larocque, and Catherine McBride, 1989. *Highlights of Aboriginal Conditions 1981-2001 Demographic Trends. Part I*. Quantitative Analysis & Socio-demographic Research Working Paper Series 89-1. Ottawa: Department of Indian and Northern Affairs.

Hamelin, Louis-Edmond, 1979. *Contribution to the Northwest Territories Population Studies, 1961-1985*. Research Paper No. 1, Science Advisory Board of the Northwest Territories. Yellowknife: Department of Information.

Johnston, R.J., Derek Gregory, and David M. Smith, 1986. *The Dictionary of Human Geography*. Oxford: Blackwell.

Krolewski, A., 1973. *Northwest Territories Statistical Abstract 1973*. Ottawa: Department of Indian Affairs and Northern Development.

Maslove, Allan M. and David C. Hawkes, 1990. *Canada's North, A Profile*. 1986 Census of Canada (Catalogue 98-122). Ottawa: Minister of Supply and Services.

Newman, James L. and Gordon E. Matzke, 1984. *Population: Patterns, Dynamics, and Prospects*. Englewood Cliffs: Prentice-Hall.

Northwest Territories, Bureau of Statistics, 1990. *Northwest Territories: . . . by the numbers*. Yellowknife: Department of Culture & Communications.

————, 1990. *1989 NWT Labor Force Survey: Labor Force Activity, Education & Language*. Report No. 2. Yellowknife: Department of Culture & Communications.

————, 1990. *1989 NWT Labor Force Survey: Wage Employment & Traditional Activities*. Report No. 3. Yellowknife: Department of Culture & Communications.

Petrovich, Curt, 1990. 'The Best Reason to Come North . . . Or Is It?' *Arctic Circle* 1(1): 36-41.

Robitaille, Norbert and Robert Choinière, 1987. 'The Inuit Population of Canada: Present Situation, Future Trends', *Acta Borealia* 1/2: 25-36.

Statistics Canada, 1982. *1981 Census Dictionary*. Catalogue 99-901. Ottawa: Minister of Supply and Services.

————, 1989. *Canada Year Book 1990*. Ottawa: Minister of Supply and Services.

Stone, Thomas, 1989. 'Urbanism, Law and Public Order: A View from the Klondike', in *For Purposes of Dominion: Essays in Honour of Morris Zaslow*, edited by K.S. Coates and W.R. Morrison. North York: Captus University Publications.

Verleun, Leo J. and Brian W. Mackenzie, 1988. *Mining Potential in Northern and Southern Canada: Guidelines for Regional Development Policy*. Centre for Resource Studies. Kingston: Queen's University.

Weinstein, Jay A., 1976. *Demographic Transition and Social Change*. Morristown: General Learning Press.

Wonders, William C., 1987. 'The Changing Role and Significance of Native Peoples in Canada's Northwest Territories', *Polar Record* 23/147: 661-71.

Yukon, Bureau of Statistics, 1990a. *Labour Market Activity Survey: Occupations*, Information Sheet #17.6 90.09. Whitehorse: Bureau of Statistics.

————, 1990b. *Labour Market Activity Survey: Job Tenure*, Information Sheet #17.9-90.09. Whitehorse: Bureau of Statistics.

Chapter 5

Abele, Frances, 1988. *Gathering Strength*. Komatik Series, Number 1. Calgary: The Arctic Institute of North America of the University of Calgary.

Armstrong, Barbara and Glenn Kendall, 1990. *Mining Industry Employment Update*. Department of Energy, Mines and Resources Canada.

Berger, Thomas R., 1977. *Northern Frontier, Northern Homeland: The Report of the Mackenzie Valley Pipeline Inquiry*, vols. 1 and 2. Ottawa: Department of Supply and Services.

Bone, Robert M. and Robert J. Mahnic, 1984. 'Norman Wells: The Oil Center of the Northwest Territories', *Arctic* 37(1): 53-60.

————, 1984. *The DIAND Norman Wells Socio-Economic Monitoring Program* Report 9-84. Ottawa: Department of Indian Affairs and Northern Development.

Courchene, Thomas J., 1986. 'Avenues of Adjustment: The Transfer System and Regional Disparities', in *The Canadian Economy: A Regional Perspective*, ed. Donald J. Savoie. Toronto: Methuen.

Dacks, Gurston, 1981. *A Choice of Futures: Politics in the Canadian North*. Toronto: Methuen.

Dicken, Peter and Peter E. Lloyd, 1981. *Modern Western Society: A Geographical Perspective on Work, Home & Well-Being*. New York: Harper & Row.

Dunbar, Moira, 1966. 'The Arctic Setting', in *The Arctic Frontier*, edited by R. St J. Macdonald. Toronto: University of Toronto Press: 3-25.

Finnie, R., 1947. 'The Epic of Canol', *Canadian Geographical Journal* 34(3): 137-9.

Gilchrist, C.W., 1988. 'Roads and Highways', *The Canadian Encyclopedia*, 2nd ed. Edmonton: Hurtig.

Gill, Alison M., 1986. 'New Resource Communities: The Challenge of Meeting the Needs of Canada's Modern Frontierpersons', *Environments* 18(3): 21-34.

————, 1989. 'Experimenting with Environmental Design Research in Canada's Newest Mining Town', *Applied Geography*, 9: 177-95.

————, and Wes Shera, 1990. 'Using Social Criteria to Guide the Design of a New Community: The Case of Tumbler Ridge', *Plan Canada* 30(1): 33-42.

Government of the Northwest Territories (GWNT), Policy and Planning, 1986. *The Northwest Territories: Economic Review & Outlook*. Yellowknife: Department of Economic Development and Tourism.

————, Bureau of Statistics, 1990. *Statistics Quarterly* 12: 4.

Hladun, Helene, 1990. 'The Sixteenth Fairmont Frontier Workshop: Bring it to Market', *Arctic Petroleum Review* 13(1): 6-12.

Howlett, Karen, 1989. 'As Dams Generate Controversy, B.C. Hydro Plugs Into Private Sector', *The Globe and Mail*, 11 November 1989: B3.

Innis, Harold A., 1930. *The Fur Trade in Canada: An Introduction to Canadian Economic History*. Toronto: University of Toronto Press.

Lucas, R., 1971. *Minetown, Milltown, Railtown: Life in Canadian Communities of Single Industry*. Toronto: University of Toronto Press.

Matthews, Ralph, 1983. *The Creation of Regional Dependency*. Toronto: University of Toronto Press.

Procter, R.M., G.C. Taylor and J.A. Wade, 1984. *Oil and Natural Gas Resources of Canada, 1983*. Geological Survey of Canada, Papers 83-31. Ottawa: Minister of Supply and Services.

Rea, K.J., 1968. *The Political Economy of the Canadian North: An Interpretation of the Course of Development in the Northwestern Territories of Canada*. Toronto: University of Toronto Press.

Robertson, Gordon, 1985. *Northern Provinces: A Mistaken Goal*. Montreal: The Institute for Research on Public Policy.

Robinson, I.M., 1962. *New Industrial Towns on Canada's Resource Frontier*. Chicago: University of Chicago Press.

Rosenberg, D.M., R.A. Bodaly, R.E. Hecky, and R.W. Newbury, 1987. 'The Environmental Assessment of Hydroelectric Impoundments and Diversions in Canada', in *Canadian Aquatic Resources*, edited by M.C. Healey and R.R. Wallace. Canadian Bulletin of Fisheries and Aquatic Sciences 215, Department of Fisheries and Oceans. Ottawa: Minister of Supply and Services.

Salisbury, Richard F., 1986. *A Homeland for the Cree: Regional Development in James Bay 1971-1981*. Montreal: McGill-Queen's University Press.

Siemens. L.B., 1973. *Single-Enterprise Community Studies in Northern Canada*. Centre for Settlement Studies. Winnipeg: University of Manitoba.

Stabler, Jack C., 1968. 'Exports and Evolution: The Process of Regional Change', *Land Economics* 44 (1): 11-23.

――――, 1987. 'Fiscal Viability and the Constitutional Development of Canada's Northern Territories', *Polar Record* 23(46): 551-68.

――――, 1989. 'Dualism and Development in the Northwest Territories', *Economic Development and Cultural Change* 37 (4): 805-39.

Stelter, G. and A. Artibise, 1982. 'Canadian Resource Towns in Historic Perspective'. in *Little Communities and Big Industries*, edited by R.T. Bowles. Toronto: Butterworth: 47-60.

Shrimpton, Mark and Keith Storey, 1990. 'Long Distance Commuting Employment: Implications for Rural and Northern Development', in *Entrepreneurial and Sustainable Rural Communities*, edited by F.W. Dykeman. Sackville N.B.: Department of Geography, Mount Allison University.

Storey, Keith and Mark Shrimpton, 1989. *Impacts on Labour of Long-Distance Commuting Employment in the Canadian Mining Industry*, ISER Report No. 3. St John's: Memorial University of Newfoundland.

Waldram, James B., 1988. *As Long as the Rivers Run: Hydroelectric Development and Native Communities in Western Canada*. Winnipeg: University of Manitoba Press.

Wallace, Iain, 1987. 'The Canadian Shield: the Development of a Resource Frontier', in *Heartland and Hinterland: A Geography of Canada*, edited by L.D. McCann. 2nd ed. Scarborough: Prentice-Hall, Chapter 11.

Watkins, Melville H., 1963. 'A Staple Theory of Economic Growth', *Canadian Journal of Economics and Political Science* 29: 160-9.

————, 1977. 'The Staple Theory Revised', *Journal of Canadian Studies* 12: 3-10.

Weller, G.R., 1977. 'Hinterland Politics: The Case of Northwestern Ontario', *Canadian Journal of Political Science* 10(4): 727-54.

Williamson, J., 1965. 'Regional Inequality and the Process of National Development: A Description of Patterns', *Economic Development and Cultural Change* 13: 3-45.

Wonders, William C., 1981. 'Northern Resources Development', in *Canadian Resource Policies: Problems and Prospects*, edited by Bruce Mitchell and W.R. Derrick Sewell. Toronto: Methuen.

Usher, Peter J., 1987. 'The North: One Land, Two Ways of Life', in *Heartland and Hinterland: A Geography of Canada*, edited by L.D. McCann. Scarborough: Prentice-Hall Canada Inc.

Zaslow, Morris, 1988. *The Northward Expansion of Canada: 1914-1967*. The Canadian Centenary Series. Toronto: McClelland and Stewart.

Chapter 6

Blair, S. Robert and Shirley G.E. Carr, 1981. *Major Canadian Projects: Major Canadian Opportunities: A Report by the Consultative Task Force on Industrial and Regional Benefits from Major Canadian Projects*. No publisher or place.

Bone, Robert M. and Robert J. Mahnic, 1984. 'Norman Wells: The Oil Center of the Northwest Territories', *Arctic* 37 (1): 53-60.

Bone, Robert M., 1984. *The DIAND Norman Wells Socio-Economic Monitoring Program*. Report 9-84. Ottawa: Department of Indian Affairs and Northern Development.

————, 1988. 'Cultural Persistence and Country Food: The Case of the Norman Wells Project', *The Western Canadian Anthropologist* 5: 61-79.

Brown, Dick, 1989. 'The OSLO Factor', *Imperial Oil Review* Summer 1989: 6-9.

Dacks, Gurston, 1981. 'The Economic Future: Non-Renewable Resources', in *A Choice of Futures: Politics in the Canadian North*. Toronto: Methuen.

Esso, 1980. *Norman Wells Oilfield Expansion: Development Plan*. Calgary: Esso Resources Canada Ltd.

Depape, Denis and Rosemary Cairns, 1985. *PCC Wrap-Up Review of Norman*

Wells Project: Proceedings and Papers. Yellowknife: Department of Indian Affairs and Northern Development.

Duffy, Patrick, 1981. *Norman Wells Oilfield Development and Pipeline Project: Report of the Environmental Assessment Panel*. Ottawa: Federal Environmental Assessment Review Office.

Fisher, Mathew, 1988. 'Ottawa, Alberta to Spend $75 Million on Mill', *The Globe and Mail*, 9 February 1988.

Green, Milford B. and David A. Stewart, 1986. *Community Profiles of Socio-Economic Change, 1982-1985*. Report 9-85. Ottawa: Department of Indian Affairs and Northern Development.

Ironside, R.G. and I. Mellor, 1978. 'The Incidence Multiplier Impact of a Regional Development Programme: A Frontier Example', *The Canadian Geographer* 22 (3): 225-51.

——— and Wolfgang Fieguth, 1990. 'The Alberta Forest Products Industry: Top-Down Initiatives Bottom-Up Problems'. Paper presented at the 1990 Annual Meeting of the Canadian Association of Geographers, Edmonton.

McKenna, Barrie, 1990. 'Hydro-Quebec Power Hikes Threaten Industry Group', *The Globe and Mail*, 27 March 1990.

Rea, K.J., 1968. *The Political Economy of the Canadian North: An Interpretation of the Course of Development in the Northern Territories of Canada to the Early 1960s*. Toronto: University of Toronto Press.

Stewart, David A. and Robert M. Bone, 1986. *Norman Wells Socio-Economic Monitoring Program: Summary Report*. Report 1-86. Ottawa: Department of Indian Affairs and Northern Development.

Stirling, Claire, 1989. 'Pulp Mills In Alberta's Boreal Forest', *Environmental Views* 12(2): 15-19.

Verleun, Leo J. and Brian W. Mackenzie, 1988. *Mining Potential in Northern and Southern Canada: Guidelines for Regional Development Policy*. Centre for Resource Studies. Kingston: Queen's University.

Chapter 7

Berkes, Fikret, 1982. 'Preliminary Impacts of the James Bay Hydroelectric Project, Quebec, on Estuarine Fish and Fisheries', *Arctic* 35(4): 524-30.

Berger, Thomas R., 1977. Northern Frontier, *Northern Homeland: The Report of the Mackenzie Valley Pipeline Inquiry*, vols. 1 and 2. Ottawa: Department of Supply and Services.

Bidleman, T.F. and others, 1989. 'Toxaphene and Other Organochlorines in Arctic Ocean Fauna: Evidence for Atmospheric Delivery', *Arctic* 42(4): 307-13.

Burnett, J.A., C.T. Dauphiné Jr, S.H. McCrindle, and T. Mosquin, 1989. *On the Brink: Endangered Species in Canada*. Saskatoon: Western Producer Prairie Books.

Cernetig, Miro, 1990. 'Cadmium, Mercury Found in Flesh of Arctic Whales', *The Globe and Mail*, 26 May 1990: A1 and A6.

Chenard, M. Paul, 1990. 'Global Atmospheric Change', in *Proceedings of the 4th Conference on Toxic Substances*. Environment Canada. Ottawa: Minister of Supply and Services.

Cloutier, Luce, 1987. 'Quand le mercure s'élève trop haut à la Baie James . . .', *Acta Borealia* 4(1/2): 5-23.

Crampton, Colin B., 1988. 'Terrain Evaluation and Pipeline Construction in the Canadian North', *The Musk-Ox* 36: 19-28.

Davies, John A., 1985. 'Carbon Dioxide and Climate: A Review', *The Canadian Geographer* 29(1): 74-85.

Duffy, Patrick, 1981. *Norman Wells Oilfield Development and Pipeline Project. Report of the Environmental Assessment Panel*. Ottawa: Minister of Supply and Services.

Dufour, Jules, 1991. 'Toward Sustainable Development of Canada's Forests', in *Resource Management and Development*, Bruce Mitchell, ed. Toronto: Oxford University Press, ch. 4.

Environment Canada, 1990. *Detailed Review Comments on the Windy Craggy Stage I Environmental and Socioeconomic Impact Assessment* (January 1990). Ottawa: Environment Canada.

French, H.M., 1976. *The Periglacial Environment*. London: Longman.

Geddes Resources Limited, 1990. *Windy Craggy Project: Stage 1: Environmental and Socioeconomic Impact Assessment*. Volume I *Project Description* and Volume 2 *Environmental Baseline Studies*. Toronto: Geddes Resources Limited.

Gill, Don and Alan D. Cooke, 1975. 'Hydroelectric Developments in Northern Canada: A Comparison with the Churchill River Project in Saskatchewan', *The Musk-Ox* 15: 53-6.

Gorrie, Peter, 1990. 'The James Bay Power Project', *Canadian Geographic*, 110(1): 21-31.

Gould, Peter, 1988. 'Tracing Chernobyl's Fallout', *Earth and Mineral Sciences* 57(4): 57-65.

Grove, Jean M., 1988. *The Little Ice Age*. London: Methuen.

Hare, F. Kenneth, 1973. 'The Atmospheric Environment of the Canadian North', in *Arctic Alternatives*, edited by D.H. Pimlott, K.M. Vincent, and C.E. McKnight. Ottawa: Canadian Arctic Resources Committee.

James Bay Energy Corporation, 1988. *La Grande Rivière: A Development in Accord with its Environment*. Montreal: James Bay Energy Corporation.

Kabzems, Alfred and D.L. Bernier, 1975. *Forestry. Final Report 4, Churchill River Study*. Saskatoon: Churchill River Study.

Kierans, T.W., 1984. *The Great Recycling and Northern Development (Grand) Canal*. Notes for a submission to the Inquiry on Federal Water Policy in Canada. St John's, Newfoundland, 24 September 1984 (available from Environment Canada, Ottawa).

Lamb, H.H., 1977. *Climate: Present, Past and Future*. 2 vols. London: Methuen.

Marsh, William M., 1987. *Earthscape: A Physical Geography*. Toronto: Wiley.

Maxwell, B., 1987. 'Atmospheric and Climatic Change in the Canadian Arctic: Causes, Effects, and Impacts', *Northern Perspectives* 18(5): 2-6.

MacInnes, K.L., M.M. Burgess, D.G. Harry, and T.H.W. Baker, 1990. *Permafrost and Terrain Research and Monitoring: Norman Wells Pipeline*. Environmental Studies No. 64, vol. II. Ottawa: Minister of Supply and Services.

Nikiforuk, Andrew and Ed Struzik, 1989. 'The Great Forest Sell-Off', *Report on Business Magazine* November 1989.

Page, Robert, 1986. *Northern Development: The Canadian Dilemma*. Toronto: McClelland and Stewart.

Porter, S.C., 1986. 'Pattern and Forcing of Northern Hemisphere Glacier Variations During the Last Millennium', *Quaternary Research* 26: 27-48.

Ripley, E.A., 1987. 'Climatic Change and the Hydrological Regime', in *Canadian Aquatic Resources*, edited by M.C. Healey. Department of Fisheries and Oceans. Ottawa: Minister of Supply and Services.

Robinson, Allan, 1991. 'Exploring the Risks at Windy Craggy', *The Globe and Mail*, 19 January: B1.

Rosenberg, D.M., R.A. Bodaly, R.E. Hecky, and R.W. Newbury, 1987. 'The Environmental Assessment of Hydroelectric Impoundments and Diversions in Canada', in *Canadian Aquatic Resources*, edited by M.C. Healey and R.R. Wallace. Ottawa: Department of Fisheries and Oceans: 71-104.

Saunders, Alan, ed. 1990. 'Pulp and Paper on the Athabasca: Economic Diversity vs. Environmental Disaster', *Northern Perspectives* 18(1): 1-12.

Searle, Rick, 1991. 'Journey to the Ice Age: Rafting the Tatshenshini, North America's Wildest and Most Endangered River', *Equinox* 55(1): 24-35.

Shkilnyk, Anastasia M., 1985. *A Poison Stronger than Love: The Destruction of an Ojibwa Community*. New Haven: Yale University Press.

Sinclair, William F., 1990. *Controlling Pollution from Canadian Pulp and Paper Manufacturers: A Federal Perspective*. Environment Canada. Ottawa: Minister of Supply and Services.

Sugden, David, 1982. *Arctic and Antarctic: A Modern Geographical Synthesis*. Oxford: Blackwell.

Tener, John S., 1984. *Beaufort Sea Hydrocarbon Production and Transportation Proposal: Final Report of the Environmental Assessment Panel.* Federal Environmental and Review Process. Ottawa: Minister of Supply and Services.

Waiten, Cathy M., 1981. *A Guide to Social Impact Assessment.* Research Branch, Corporate Policy, Department of Indian and Northern Affairs. Ottawa: Department of Indian and Northern Affairs.

Williams, Peter J., 1979. *Pipelines and Permafrost: Physical Geography and Development in the Circumpolar North.* London: Longman.

———, 1986. *Pipelines & Permafrost: Science in a Cold Climate.* Ottawa: Carleton University Press.

Chapter 8

Bone, Robert M., 1973. 'The Number of Eskimos: An Arctic Enigma', *Polar Record* 16 (103): 553-7.

——— and Milford B. Green, 1983. 'Housing Assistance and Maintenance for the Métis in Northern Saskatchewan', *Canadian Public Policy*, 9 (4): 476-86.

Brown, Jennifer S.H., 1988. 'Métis', *The Canadian Encyclopedia*, James H. Marsh, ed. Edmonton: Hurtig Publishers.

Canada, National Health and Welfare, 1968 to 1987. *Report on Health Conditions in the Northwest Territories.* Yellowknife: Medical Services.

Canada, Department of Indian and Northern Affairs, 1989. *Northern Indicators.* Ottawa: Strategic Planning Directorate.

Damas, David, ed., 1984. 'Arctic'. *Handbook of North American Indians*, William C. Sturtevant (general editor). Washington: Smithsonian Institution.

Dickman, Phil, 1969. 'Thoughts on Relocation', *The Musk-Ox* 6: 21-31.

Dorais, Louis-Jacques, 1989. 'Bilingualism and Diglossia in the Canadian Eastern Arctic', *Arctic* 42(3): 199-207.

Driver, Harold E., 1961. *Indians of North America.* Chicago: University of Chicago Press.

Freeman, Minnie A., 1988. 'Inuit', *The Canadian Encyclopedia*, James H. Marsh, ed. Edmonton: Hurtig Publishers.

Frideres, James S., 1983. *Native People in Canada: Contemporary Conflicts.* Scarborough:Prentice-Hall.

Gamble, Donald J., 1986. 'Crushing of Cultures: Western Applied Science in Northern Societies', *Arctic* 39(1): 20-3.

Hagey, N. Janet, Gilles Larocque, and Catherine McBride, 1989. *Highlights of Aboriginal Conditions 1981-2001: Part I Demographic Trends; Part II Social Conditions; and Part III Economic Conditions.* Quantitative Analysis & Socio-demographic Research Working Paper Series 89-1. Ottawa: Department of Indian and Northern Affairs Canada.

Helm, June, ed., 1981. 'Subarctic', *Handbook of North American Indians*, William C. Sturtevant (general editor). Washington: Smithsonian Institution.

Hodgson, Gordon W., ed., 1986. 'Languages of the Majority', *Information North*. The Arctic Institute of North America. Summer Issue. Calgary: The University of Calgary.

Irwin, Colin, 1988. *Lords of the Arctic: Wards of the State, Inungnut*. Rankin Inlet: Keewatin Inuit Association.

Jenness, Diamond, 1932. *Indians of Canada*. Geological Survey Museum Bulletin: Anthropology Series No. 15. Ottawa: Queen's Printer.

Keenleyside, Anne, 1990. 'Euro-American Whaling in the Canadian Arctic: Its Effects on Eskimo Health', *Arctic Anthropology* 27 (1): 1-19.

Klausner, Samuel Z. and Edward F. Foulks, 1982. *Eskimo Capitalists: Oil, Politics, and Alcohol*. Totowa: Allanheld, Osnum & Co.

Kroeber, Alfred L., 1939. *Cultural and Natural Areas of Native North America*. University of California Publications in American Archaelogy and Ethnology, Vol. 38. Berkeley: University of California Press.

McCue, Harvey, 1988. 'Indian', *The Canadian Encyclopedia*, James H. Marsh, ed. Edmonton: Hurtig Publishers.

McGhee, Robert, 1984. 'Thule Prehistory of Canada', in *The Arctic*, ed. David Damas. Handbook of North American Indians, Vol. 5. Washington: Smithsonian Institution.

McMillan, Alan D., 1988. *Native Peoples and Cultures of Canada: An Anthropological Overview*. Vancouver: Douglas & McIntyre.

McNabb, Steven, 1990. 'Native Health Status and Native Health Policy: Current Dilemmas at the Federal Level', *Arctic Anthropology* 27(1): 20-35.

Mooney, James, 1928. 'The Aboriginal Population of America North of Mexico'. *Smithsonian Miscellaneous Collections*, 80: 7.

Robitaille, Norbert and Robert Choinière, 1987. 'The Inuit Population of Canada: Present Situation, Future Trends', *Acta Borealia* 4, (1-2): 25-36.

Salisbury, Richard F., 1986. *A Homeland for the Cree: Regional Development James Bay, 1971-1981*. Kingston: McGill-Queen's University Press.

Shkilnyk, Anastasia, M., 1985. *A Poison Stronger than Love: The Destruction of an Ojibwa Community*. New Haven: Yale University Press.

Smith, Derek G., 1984. 'Mackenzie Delta Eskimos', in *Handbook of North American Indians*, William C. Sturtevant, ed. Vol. 15 (Arctic). Washington: Smithsonian Institution.

Stenbaek, Marianne, 1987. 'Forty Years of Cultural Change Among the Inuit In Alaska, Canada and Greenland: Some Reflections', *Arctic* 40 (4): 300-9.

Stewart, David A., 1987. *The Indicator Approach in the Examination of Spatial Variations in the Level of Development of Natives and the Concept of Dualism in Canada*. M.A. Thesis. Saskatoon: Department of Geography, University of Saskatchewan.

Williamson, Robert, 1974. *Eskimos Underground: Socio-Cultural Change in the Canadian Central Arctic*. Uppsala: Almqvist and Wiksell.

Wonders, William C., 1987. 'The Changing Role and Significance of Native Peoples in Canada's Northwest Territories', *Polar Record* 23 (147): 661-71.

Chapter 9

Asch, Michael, 1984. *Home and Native Land: Aboriginal Rights and the Canadian Constitution*. Toronto: Methuen.

Berger, Thomas R., 1977. *Northern Frontier, Northern Homeland: The Report of the Mackenzie Valley Pipeline Inquiry*, vols. 1 and 2. Ottawa: Department of Supply and Services.

Beveridge, John, 1979. 'The Rabbit Lake Commuting Operation: A Case for Mutual Adaptation', in *Proceedings: Conferences on Commuting and Northern Development*, Michelle Mougeot ed. Saskatoon: Institute for Northern Studies.

Bone, Robert M. and Milford B. Green, 1984. 'The Northern Native Labour Force: A Disadvantaged Work Force', *The Operational Geographer* 3: 12-14.

Dacks, Gurston, 1981. *A Choice of Futures: Politics in the Canadian North*. Toronto: Methuen.

Duerden, Frank, 1990. 'The Geographer and Land Claims: A Critical Appraisal', *The Operational Geographer* 8(2): 35-7.

Government of Canada, Indian and Northern Affairs, 1985. *The Western Arctic Claim: The Inuvialuit Final Agreement*. Ottawa: Department of Indian Affairs and Northern Development.

Government of the Northwest Territories, Department of Economic Development and Tourism, 1986. *The Northwest Territories: Economic Review & Outlook*. Yellowknife: Department of Culture & Communications.

Government of Northwest Territories, Bureau of Statistics, 1990. *Statistics Quarterly*. Yellowknife: Department of Culture & Communications.

———, 1990. *1989 NWT Labour Force Survey: Wage Employment & Traditional Activities*. Report No. 3. Yellowknife: Department of Culture & Communications.

Hawkes, David C. (ed.), 1989. *Aboriginal Peoples and Government Responsibility*. Ottawa: Carleton University Press.

Hobart, Charles, 1989. 'Company Town or Commuting: Implications for Native People', in *The Commuting Alternative*, Robert Rubson ed. Winnipeg: Institute of Urban Studies.

Larocque, Gilles Y. and R. Pierre Gauvin, 1989. *Basic Departmental Data - 1989.* Quantitative Analysis and Socio-demographic Research. Ottawa: Department of Indian Affairs and Northern Development.

———, 1990. *Basic Departmental Data - 1990.* Quantitative Analysis and Socio-demographic Research. Ottawa: Department of Indian Affairs and Northern Development.

LaRusic, I.E., 1982. *Income Security of Subsistence Hunters: A Review of the First Five Years of the Operation of the Income Security Program for Cree Hunters and Trappers.* Corporate Policy. Ottawa: Department of Indian Affairs and Northern Development.

Miller, J.R., 1989. *Skyscrapers Hide the Heavens: A History of Indian-White Relations in Canada.* Toronto: University of Toronto Press.

Peters, Evelyn J., 1989. 'Federal and Provincial Responsibilities for the Cree, Naskap and Inuit under the James Bay and Northern Quebec, and Northeastern Quebec Agreements', in *Aboriginal Peoples and Government Responsibility: Exploring Federal and Provincial Roles* ed. D.C. Hawkes. Chapter 6. Ottawa: Carleton University Press.

Robinson, Michael and Elmer Ghostkeeper, 1987. 'Native and Local Economies: A Consideration of Economic Evolution and the Next Economy', *Arctic* 40(2): 138-44.

———, 1988. 'Implementing the Next Economy in a Unified Context: A Case Study of the Paddle Prairie Mall Corporation', *Arctic* 41(3): 173-82.

Robinson, Michael, Michael Pretes and Wanada Wuttunee, 1989. 'Investment Strategies for Northern Cash Windfalls: Learning from the Alaskan Experience', *Arctic* 42(3): 265-76.

Salisbury, Richard F., 1986. *A Homeland for the Cree: Regional Development in James Bay, 1971-1981.* Montreal:McGill-Queen's University Press.

Schellenberger, Stan and John A. MacDougall, 1986. *The Fur Issue: Cultural Continuity: Economic Opportunity.* Report of the House of Commons Standing Committee on Aboriginal Affairs and Northern Development. Ottawa: Queen's Printer.

Stabler, Jack C., 1989a. 'Dualism and Development in the Northwest Territories', *Economic Development and Cultural Change*, 37(4): 805-40.

———, 1989b. 'Jobs, Leisure and Traditional Pursuits: Activities of Native Males in the Northwest Territories', *Polar Record* 25(155): 295-302.

Stevenson, D., 1968. *Problems of Eskimo Relocation for Industrial Employment.* Northern Co-ordination and Research Centre. Ottawa: Department of Northern Affairs and National Resources.

Tough, George, 1972. 'Mining in the Canadian North', Chapter 4 in *The North*, ed. William C. Wonders. Studies in Canadian Geography. Toronto: University of Toronto Press.

Usher, Peter J., 1982. 'Unfinished Business on the Frontier', *The Canadian Geographer* 26(3): 187-90.

——— and George Wenzel, 1987. 'Native Harvest Surveys and Statistics: A Critique of their Construction and Use', *Arctic* 40(2): 145-60.

——— and George Wenzel, 1989. 'Socio-Economic Aspects of Harvesting', Chapter 1 in *Keeping On the Land: A Study of the Feasibility of a Comprehensive Wildlife Harvest Support Program in the Northwest Territories* by Randy Ames, Don Axford, Peter Usher, Ed Weick, and George Wenzel. Ottawa: Canadian Arctic Resources Committee.

Weissling, Lee E., 1991. 'Inuit Life in the Eastern Canadian Arctic, 1922-1942', *The Canadian Geographer* 35(1): 59-69.

Wenzel, G.W., 1986. 'Canadian Inuit in a Mixed Economy: Thoughts on Seals, Snowmobiles, and Animal Rights', *Native Studies Review* 2(1): 69-82.

Whittington, Michael S. (co-ordinator), 1985. *The North*. Toronto: University of Toronto Press.

Williamson, Robert G. and T.W. Foster, 1975. *Eskimo Relocation in Canada*. Institute for Northern Studies. Saskatoon: University of Saskatchewan.

Wolfe, Jackie and Lynn Convery, 1989. *Microenterprises on Three Ontario Indian Reserves: Opportunities and Constraints*.

Chapter 10

Bell, Mike, 1989. *The Government of the Northwest Territories' Native Employment Policy: A Review and Assessment*. Report for the Special Committee on the Northern Economy of the Government of the Northwest Territories. Yellowknife: Inukshuk Management Consultants.

Berger, Thomas R., 1977. *Northern Frontier, Northern Homeland: The Report of the Mackenzie Valley Pipeline Inquiry*, vols 1 and 2. Ottawa: Department of Supply and Services.

Britton, John N.H., 1988. 'A Policy Prospectus on Regional Economic Development: The Implications of Technological Change', *Canadian Journal of Regional Science* 11(1): 147-65.

Canada, 1988. *The James Bay and Northern Quebec Agreement, The Northeastern Quebec Agreement: Cree-Inuit-Naskapi, Annual Report 1988*. Ottawa: Minister of Supply and Services.

Dacks, Gurston, 1986. 'The Case Against Dividing the Northwest Territories', *Canadian Public Policy* 12(1): 202-13.

Fisher, Mathew, 1988. 'PM to Sign Northern Accords on Energy, Land Claims', *The Globe and Mail* 5 September 1988: A3.

Galt, Virginia, 1991. 'Gold Mine to Give Preference to Natives', *The Globe and Mail*, 9 April 1991: A4.

Government of the Northwest Territories, 1989. *The Scone Report: Building Our Economic Future*. Yellowknife: Legislative Assembly of the Northwest Territories.

Irwin, Colin, 1988. *Lords of the Arctic: Wards of the State: The Growing Inuit Population. Arctic Resettlement and Their Effects on Social and Economic Change*. Report prepared for Health and Welfare Canada. Rankin Inlet: Keewatin Inuit Association.

Lyck, Lise, 1989. 'Greenland: Ten Years of Home Rule', *Polar Record* 25(155): 343-6.

Platiel, Rudy, 1991. 'Canada is Arctic Protection Link', *The Globe and Mail* 10 June 1991: A7.

Salisbury, Richard E., 1986. *A Homeland for the Cree: Regional Development in James Bay, 1971-1981*. Montreal: McGill-Queen's University Press.

Williams, Peter J., 1986. *Pipelines & Permafrost: Science in a Cold Climate*. Ottawa: Carleton University Press.

Index

economic structure, northern
economy, 115-34, 232
economies of scale, 137, 153
economy: community development,
211-12; land-based, 203; and
migration, 86-7; Native, 11-12, 103,
212-15; and trading areas, 11
education, 47, 102-103, 197, 198, 204,
215, 238
Eldorado Creek, 137
Eldorado Mining and Refining Ltd,
128
Eldorado Nuclear Ltd, 104-105
Ellesmere Island, 30, 32
employment, 208, 222; impacts of
Norman Wells project, 151-2;
income, 205-208; in resource
industries, 106, 232
English River (Ontario), 168
environment: fragility of, 17; and
industrial pollution, 68, 119, 157-8,
166-70, 233; and legislation, 161,
231; and Native issues, 165; and
resource development, 158-9, 165,
231
Environment Canada, 161
environmental disasters, 166-70
Environmental Assessment: Board
(Ontario), 172; Panel, 163
Environmental Assessment and
Review (EARP), 162, 166
Environmental Contaminants Act,
161, 178
Environmental Impact Statement
(EIS), 163
Environmental and Review Office,
141
environmental and social assessment
process, 81-2
epidemics, see disease
equalization payments and territorial
governments, 112
eskers, 30-1
Esso Resources Canada Ltd, 106,
142, 146, 148, 163, 165, 210
European influence, and Indian
economies, 46, 48
export markets, 12-13, 99-100, 104,
106, 110, 119, 121, 168, 240

Extreme North zone, 8-9

fall-out of radioactive debris, 174-5
Family Allowance Payments, 67, 207
Far North zone, 8-9
Faro (Yukon), 100, 105, 121
Federal Environment and Review
Office (FEARO), 161-3, 178
Federal Environmental and Assessment: Office, 150; Process, 163
Federation of Ontario Naturalists,
173
fisheries, 169
fishing, see hunting
Flin Flon (copper smelter), 116, 138
fly-in systems, see air commuting
Food and Agriculture Organization,
78
food chain, 158, 160, 169, 231
food (Native), see country food
food supply, 41-2, 183, 187; and fur
traders, 49-50, 56
Foothills Pipe Lines Ltd (Foothills),
141, 165
Foreign Investment Review Act
(FIRA), 112
foreign ownership, resource industry,
110, 138, 140-1; see also Japan;
United States
forest industry and the environment,
171-4, 231
Forest Management Licence Area,
141
forest parkland subzone, 21
forest resource, 56, 118-19, 141
Forests For Tomorrow, 172-3
Formula Financing (1985, 1989), 113
Forts: Chipewyan, 49, 50; Churchill,
53; Franklin, 153; Good Hope, 153;
McMurray, 82; Norman, 153, 155;
Providence, 153; St John, 101;
Simpson, 23, 153, 155; Smith, 153
fossil fuels, 174
Fourth World, 5
Franklin, Sir John, 54
Free Trade Agreement (1989), 12-13,
110, 238
French traders, 49
Frobisher, Martin, 44